Early

TOO YOUNG TO BE OLD

"Too Young to be Old is one of those books that when you see what percentage you have read, you are glad you have a long way to go. It was bittersweet and heartfelt as Frank learned to work with the old folks."
 - Patricia Steele, author of *Wine, Vines and Picasso*

"Funny, intriguing, alarming, mystifying, and enlightening, this is also a superb lead-in to Frank's travel memoirs proper."
 - Terry Murphy, author of *Weekend in Weighton*

"Honest, whimsical, hilarious and keenly observing - a superb first course!"
 - Grady Harp, Top 100 Amazon reviewer

"You won't go wrong with this book, or frankly any book by Mr. Kusy. This may be my favorite, but it doesn't take away from all the others- it just enriches his total opera omnia."
 - Jayna Mittener, Top 1000 Amazon reviewer

"Frank is a wonderful story teller who effortlessly combines a terrific sense of humour and bags of charm in his writing – I really warmed to him."
 - Mrs BEH Towers, Amazon reviewer

"Great stuff from a warm and witty storyteller."
 - Mark Roman, author of *The Ultimate Inferior Beings*

"Well Frank, you've done it again - written a funny and almost too good to be true memoir."
 - Joann R. Greene, Amazon reviewer

Also by Frank Kusy

Kevin and I in India
Rupee Millionaires
Off The Beaten Track
He ain't Heavy, He's my Buddha
Ginger the Gangster Cat
Ginger the Buddha Cat

All titles available from Grinning Bandit Books

Too Young to be Old: From Clapham to Kathmandu

Frank Kusy

First published in 2015 by
Grinning Bandit Books
http://grinningbandit.webnode.com

© Frank Kusy 2015

'Too Young to be Old: From Clapham to Kathmandu' is the copyright of Frank Kusy, 2015.

The people and events in this book are portrayed as perceived and experienced by Frank Kusy. Some names have been changed for privacy reasons.

All rights reserved.
No part of this book may be reproduced or transmitted in any form or by any means, electronic, digital or mechanical, without permission in writing from the copyright owner.

ISBN 978-0-9934047-0-2

Cover design by Amygdaladesign

DEDICATION

For Julie Haigh, a tireless supporter of memoir writers.

Contents

Author's Note	v
1. Old Bill	1
2. Fire Drill	7
3. My Grandfather	16
4. Pocket Money Day	27
5. In the Hot Seat	37
6. I knew Elizabeth Taylor	46
7. A Contract is a Contract	59
8. Don't put your Director on the Stage	69
9. Bertie Steps Out	80
10. Phoney Pheasants and Plasticine Pigs	94
11. Devils and Bingo Balls	101
12. Anna	111
13. The Importance of Mission	123
14. Too Young to be Old	132
15. Kevin and I in India	143
16. From Tozo to Tozan	154
17. Going Japanese	165
18. Birth of a Travel Writer	177
Postscript	182
A Message from the Author	183
Acknowledgements	185
About the author	186

Author's Note

I would like to say it was the search for spiritual meaning in my life, or even a noble desire for world peace, that brought me to Buddhism. But no, it was a wild-eyed, crazy old gentleman known as Old Bill. I would also like to say that it was a natural talent for travel and adventure that propelled me into 30 years of globetrotting and travel writing. But no, it was another crazy, eccentric old gentleman called Bertie. This is the story of how these two venerables and two years working at an old people's home in South London changed my life forever.

Chapter 1

Old Bill

My first impression of the Edwin Maudsley home was not a good one.

'You're late!' growled Old Bill as he answered the door. 'Should've been here half an hour ago.'

I checked my watch and protested I was actually a few minutes early.

Old Bill looked affronted by my impertinence. Then he peered short-sightedly at me and bristled at what he saw. 'You're not the Mayor of Lambeth!'

I could only agree.

'You're an imposter!' he accused before trying to whack me over the head with his white stick.

I could see Old Bill and I were going to get on famously.

Taking a step back, out of range of the stick, I wondered what I had got myself into. It was November 1982 and I was standing here, outside the tall, red-brick building, to please my mother, who was determined that I succeed in what she called a 'proper' job. I had tried and failed in insurance. I had tried and failed in publishing. Indeed, I had tried and failed in eleven different jobs since leaving university. If I were to fail here, in the relatively modest arena of elderly care, my mother would give up on me completely. Little did I know the experience would push me into a journey of inner discovery that would soon lead me to India.

For now, though, things had not begun well. Fortunately, a tall, broad-shouldered figure in a pink dress, and a pile of perfectly coiffured white hair, appeared behind Old Bill to disarm him.

'I'm Mrs Butterworth, the Matron,' she said above Old Bill's protests as she escorted me to my office. 'Don't worry about him. Manic depressive. Gets these crazes. Feels the need to take "responsibility" for something. First it was the visitor's room, then it was the bird bath, and now he's got it into his head he's our doorman. He won't let anyone in unless he likes them, and since he's half-blind and can't see who's at the door, he won't let them in anyway.'

I nodded sympathetically, and sized Matron up. It wasn't just her dress that was pink, *everything* about her was pink – from her brightly coloured slippers right up to her luminous pink watchstrap, nails and lipstick. But with a mis-matching magenta jacket over the rest of the ensemble, she looked a bit like Barbara Cartland without the dress sense.

'Are you *expecting* the Mayor of Lambeth?' I asked.

'No,' sniffed Matron. 'Last time he visited, Bill mistook him for the man from the Social and demanded his pocket money – so I doubt we'll be seeing him any time soon. Nevertheless, Old Bill's hoping he'll come back with his Christmas bonus.'

The incumbent Administrator, Mrs Gregory, was waiting for me in the office. A bubbly little dumpling with a large bosom and an enormous mop of grey hair, she closed the door behind me and said: 'I see you've met that terrible woman.'

I was somewhat taken aback by this strange greeting. 'You mean Matron?'

'Oh, I'm sorry, I'm not being very welcoming, am I? But I can't help it. I hate her.'

I settled into a spare chair, noticing as I did so that the office was surprisingly bare: just a desk with a typewriter on it, a single filing cabinet, and a steel safe in the corner.

'Well, she was very nice to me,' I responded. 'Is she really that bad?'

Mrs Gregory squeezed her generous hips into the other chair and started nodding at me fiercely. 'Yes, she is. She's evil. We were supposed to organise a bric-a-brac sale together but on the morning before the big day she accosted me in the lift and said, "I can't do it! I'm not doing it! It's too short notice! Why should we do it? We're not getting paid for it!" Then she swanned off, leaving me to do everything myself. To top it all, at the actual event, she barged in, having had a rethink, and stopped the proceedings to give a speech in which she took all the credit for organising it. I was fuming!'

'Not very collegiate,' I said. 'Did you have a word with her afterwards?'

'You bet I did. Two days later, she knocked at my door, smiling sweetly, and said: "I guess I lost it a bit back there, I'm sorry if I seemed rude." Well, I wasn't having that. "Do not ever speak to me again," I said. "To me, you are dead." And shut the door in her face.'

Over the next hour or two, with Mrs Gregory chattering madly at my shoulder, I began learning my new job. It seemed I was to be clerical officer, administrator, typist, fund-raiser, personnel officer, committee secretary and general dogsbody all rolled into one. I began kicking myself for taking on a job with so much responsibility – responsibility was *not* my middle name.

Then, just as abruptly as Mrs Gregory had welcomed me, she left. 'Right, that's it. I'm off,' she said, struggling out of her seat and into her coat. 'Here. Here are the keys. Anything else you need to know, just call me at home.'

I was speechless. 'You're going already? Aren't you coming back?'

'No, I'm not. If I have to spend one more minute in the same building as that woman, I'm going to explode.' As she reached the door, she whispered, 'Don't let her get near you. She's *poison.*'

I blinked in surprise.

'Her *aura*,' she explained. 'Don't let her get into your *aura.*' Then, tapping the side of her nose, she was off.

I stared at the closed door. Struggling to remember even a tenth of her rapid machine-gun instructions, I felt like a toddler in a bathtub thrown into a tsunami – without a paddle.

Moments later, there was a tentative tap on the door, and Old Bill hobbled in.

'What do you want for lunch, Mr Queasy?' he quavered. 'There's quiche or bolognaise.'

For a moment I debated the wisdom of placing an order with Old Bill, but perhaps I had become his 'responsibility' for the day. 'I'll have the quiche, Bill, if that's not too much trouble.'

'No trouble at all.' Bill turned and staggered out, to be replaced by the deputy Matron. A youngish girl in her mid-20s with the weight of the world on her martyred shoulders. Her tolerance of elderly antics had evidently worn thin.

'Oh, he's a pain, that one. Don't expect your lunch before suppertime. And that's only if he makes it to the kitchen. Blind as a bat. Not as blind as Mr Goodall, mind. Three weeks it took me to teach him to feel his way round the home by himself. And then the old bugger went and died in the toilet. What a waste of time!'

My eyebrows raised and stayed on hoist. I'd never heard elderly people referred to with such contempt. How could these people help old folk if they didn't like or respect them?

As I was about to pose this question an angry scuffle broke out in the corridor outside.

'Oh Gawd, what now?' sighed the deputy. 'I bet it's that bloody Bill again.'

I followed her out of the office and saw two elderly men sword-fighting with white sticks.

'Come on, then!' Bill challenged his opponent. 'You come here, and I'll kick your arse full of old boots and leave it there like a stamp on a letter!'

The other guy, a tall Irishman with a manic gleam in his eye, was brandishing a plate with a potato on it. 'Will you look at that!' he brayed in a thick brogue. 'How am I expected to eat this muck? It's got oil all over it. And that's the man, that's the evil creature, who put it there!'

A thrill of excitement ran through the main lounge, and twenty pairs of aged eyes swivelled round to catch a rare glimpse of action. Knitting needles clattered to the floor, dog-eared books slumped into laps, and two old ladies in the corner started laying bets.

Just then, Matron turned up. 'Now, what's all this?' she screeched in a shrill tone. 'Oh, I should have known. It's you, Bill, isn't it?'

Old Bill took one look at her, and promptly deflated. 'Who is it? Who is it? Can I have a cup of tea?'

Matron fixed him with a withering glance, and deprived him of his stick again. For a moment, I thought she was going to cleave him in two with it, but then he defused the situation by collapsing on her shoulder and murmuring: 'Where do I live? Can I book a room for the night?'

It was a pantomime, even I knew that, but all my alarm bells were ringing.

If Old Bill needed this much attention, and if Matron was as unbalanced as she looked and sounded, I was going to need help.

Right now.

Chapter 2

Fire Drill

To this day, I don't know why I reached out to Brenda. She had looked after me when she'd been my editor boss at the Financial Times, and had nudged me awake quite a few times when I'd dozed off at my desk, but she wasn't a particularly close friend. I guess, if anything, it was the change I'd seen in her after she'd converted to Buddhism that made me pick up the phone.

'I'm so sorry to bother you,' I said, 'but I need some advice and don't know who else to ask.'

'Shoot,' said Brenda in that familiar deep, throaty voice of hers. 'Anything to oblige a pal.'

'Well, I think I'm going mad. I've only been on this job a few hours and I'm surrounded by nutters. Unless you've got any bright ideas, I'm out the door the moment we finish this conversation.'

Brenda laughed. 'I know you, Frank. The first sign of trouble and you want to cut and run. Look, if the lunatics are running the asylum, why don't you chant to become the head lunatic?'

'Chant to do what?'

'Chant to be in charge of the place. What have you got to lose?'

Well, the answer to that was: 'nothing'. If I walked out that door, I would merely trudge back to my dreary little bedsit in Clapham and chalk up a round dozen of failed employments. Moreover, money was tight and nobody would be getting any cards for Christmas, let alone a present.

'Okaaay,' I said in a cautious tone. 'So what is this chant, then? Because if it's that "Oh money, give me some" one, I already tried it and it didn't work.'

Brenda laughed again. 'You're thinking "Om Mani Padme Hum", silly! That's the mantra or prayer of Tibetan Buddhists. What we do is a Japanese chant: Nam Myoho Renge Kyo. It means "I devote myself to the Creative Force of Life" and is pronounced Nam Mee-o Ho Ren-gay K'yo. I gave you a card about it, remember?'

I did remember. She'd given it me a few months before, and I'd put it in my back pocket. It was probably still there.

'Well, find that card,' she said, 'and try it out. Find a spot on the wall to concentrate on, put your hands together, and sit, kneel, stand on your head, it doesn't matter, just say that chant – over and over, the same four words – to not only create maximum value in your work situation, but to be given the authority to transform the lives of everyone in it.'

'By tomorrow?'

'No, silly. Buddhism is reason. Give it a month. Do you think you can manage that?'

I wasn't sure, but then, as the phone clicked off, something happened to make my mind up.

The doors to the home burst open and a short, red-faced madman blew in like a truculent tornado.

'I've just about had it!' he raged at nobody in particular. 'I put my mother in here to be taken care of, and I just had a phone call saying she's sitting in her chair crying into her porridge. Who's looking after her? Nobody. Where's that bloody Matron?'

I regarded the crazed figure with astonishment, along with an inordinate amount of fear. He reminded me of my stepfather Bert, who was also prone to sudden, inexplicable fits of temper. Indeed, without the horn-rimmed spectacles and the bushy eyebrows flaring above them, he could have passed for Bert's evil twin.

'Who's *he?*' I asked a passing assistant. She shrugged, saying, 'That's Mr Parker, the home's Chairman. Best stay out of his way.'

Matron was evidently expert at staying out of his way. I saw a flash of pink in the dining area, beyond the wide lounge, and she was gone. Which suddenly left me the only staff member in sight.

'Who are you?' snarled Mr Parker, overturning a table and swinging a chair about in the air. 'And where's Mrs Gregory?'

I flinched. 'Erm...I'm Mr Kusy, the new Administrator. Mrs Gregory has gone home. I don't think she's coming back.'

The chair dangled dangerously above my head. 'Well, this is shit in a biscuit!' declared Mr Parker. 'We got a fire drill tomorrow! Who's going to organise that?'

'Matron?'

'Matron? Don't make me laugh. She couldn't organise a piss-up in a brewery. No, I'll have to do it myself, won't I? And she and her cronies better be up to the mark, or I'll sack the whole sodding lot of them!'

I watched open-mouthed as my new employer tossed the chair carelessly over his shoulder and strutted off in search of his mother. Then I reached into my back pocket and retrieved

Brenda's card before retiring to my office and chanting my socks off.

*

I don't know whether it was the chanting, but the next day went better. Well, for me it did, though it was a black day for everybody else – particularly John Trundle.

John Trundle was a stout young care assistant and none too bright. He knew to avoid Mr Parker, and he knew to defer to Matron, but was that was about it. When the hot-headed Chairman arrived to organise the fire drill, John made a beeline for the nearest toilet, locked himself in and missed the all-important briefing.

'Okay, you lot,' said Mr Parker. 'We're going to have a fire drill. See this big bit of bright red metallic paper with flames licking out of it? Well, that's the fire sign. I'm going to go hide it now, and when the alarm bell goes off, you've got to go find it…and bring it back here.'

Two minutes passed as Matron and me and the other officers waited in my office, and then he was back. 'Right, off you go!' We scattered.

I honestly didn't know where the fire was going to be, but then the bell went off and I happened to go into the TV room and spotted this red thing sitting in the waste paper bin, hidden behind a chair. 'Hello,' I thought excitedly. 'This is a feather in my cap!'

But my triumph was short lived. At that very moment, John Trundle emerged from the toilet and the Chairman spotted him.

'Oi, you, Trundle!' he bellowed. 'What you doing in there? Why aren't you looking for the fire?'

John stopped dead in his tracks and looked dimly back at

Mr Parker. 'What fire?' he quavered, the colour draining from his face.

'What fire? Go into the TV room, and look for something which resembles a fire. Mr Kusy's just found it, now I want *you* to find it!'

John wandered into the telly room, looked around helplessly a bit, and then walked out again. 'I can't see anything,' he said. 'Nothing in there.'

'Nothing in there?' echoed Mr Parker. 'Go back and look again!'

Four times John looked for that sign and four times he came back empty-handed. 'Cum 'ere!' said Mr Parker in the end and dragged him into the room and stuck his nose in the bin with the fire notice in it. If it had been a real fire, the whole place would have been a cinder before John located it.

*

Over lunch, I got to know the deputy Matron, Miss McCann, a little better. Her bitterness at the world, and apparent lack of sympathy for the elderly, stemmed from two factors. First, she was being used as nightly target practice by her boyfriend, who was fond of lobbing empty beer bottles at her head. Second, she had had prolonged contact with Old Bill.

'I've just about had with him!' she said, stabbing the sausage in her Toad in the Hole viciously as though it might be Bill's head. 'Last night was the final straw.'

I asked what had happened.

'First, he avoided his bath by going upstairs, ringing his bell continuously, and complaining to all the assembled staff that his dirty trousers had been taken away for laundering. Then he threatened to keep ringing the emergency bell until they were returned. When this didn't work, he appeared in the dining room in his underpants and staged a hunger strike for his trousers. So I dragged the trousers out of the laundry – they were soaking wet by this time – and said, "There you go, there's your bloody trousers!" And what did he do? He started loudly bragging that he had all the staff "running up and down" for him. I could have throttled the bugger.'

The pale-faced young deputy was beside herself with rage.

'What's Bill's story, then?' I asked her. 'Why does he need so much attention?'

'His wife used to wait on him hand and foot. He nearly starved to death after she fell ill with cancer. He couldn't understand why she couldn't come downstairs to feed him. In the end, having waited three days for her to serve up his supper, he went upstairs to see what was going on. She was dead.'

'Well, that explains a lot. Where was he earlier? I didn't see him during fire drill.'

Miss McCann permitted herself a wry smile. 'We locked him in his room. There's no way we could have Bill on the loose at a fire drill. I'm not sure I'd even let him out in a *real* fire.'

I smiled. 'Have you had a real fire?'

'Oh yes. Last week the alarm went off, the fire brigade turned up, and a load of fireman appeared at Matron's door. She was inside, shampooing her hair and boiling potatoes. Actually, it was the potatoes that set the alarm off. So she appears in the doorway, with her hair in rollers, and all these firemen pour in. "I was only boiling potatoes!" she cried, but they didn't believe her.'

From her glee it was evident the put upon deputy had enjoyed seeing her pompous boss taken down a peg or two. But then she bent forward in an attitude of total seriousness and said, 'Don't get me wrong, I'm not trying to take the rise out of Matron: do that round here and you're dead. Only one thing worse than getting on the wrong side of Matron, and that's getting Parker's back up. He'll sack you for sure.'

Back in my office, I was surprised to find it full of people. First there was old Miss Johnson wanting me to address an envelope containing two chocolate digestive biscuits to a friend in Margate. Then there was Miss Morton, a social worker wanting details of the home. Standing behind her was Mr Seeger with two up-to-date wallpaper catalogues, determined to give me a complete description of his holiday in Madeira. Elderly Mr Reitz had wandered in wondering if I would pay for his Flora margarine out of petty cash, Miss Sutton had followed on to tell me why she couldn't sleep last night, and Mrs Everitt had tagged along behind her because she had nothing better to do. The developing conga line of curious residents was only broken up by Matron, who burst into my office and shepherded them all out again, barking a terse: 'They're my responsibility, dear. If they have any problems, they should come to *me!*'

Closing the door on her, I put my hands together and started chanting again. And found myself grinning from ear to ear. 'So the chaos continues,' I chuckled inside. 'How many more surprises can this day bring?'

As though in answer to the question there was a sharp knock on the door and Mr Parker strode in. 'I've been looking at you, Mr Kusy,' he said, unexpectedly shaking my hand. 'And so far, I like what I see. Call me Jack!'

I was so taken aback, I could hardly speak. 'Erm...well, thank you...err...Jack. But I haven't really done anything.'

The bushy browed tyrant gave me a macabre grin. 'Anyone who can find my fire sign in less than a minute is alright by me,' he said. 'And let me tell you right now, I'm making some big changes round 'ere shortly. You keep your nose clean and continue to show promise, and you might find yourself included in them. Got it?'

'Got it,' I said weakly as I let him out, and then my heart began thumping in my chest. Fear, excitement and trepidation all washed over me as I considered the implications of his statement.

'Blimey,' I thought. 'Is this chanting working already?'

*

As I walked home that night, trudging through the new layer of thin snow papering Clapham Common, I wondered what I was doing working with old people. I also wondered what I was doing in Clapham.

The answer to the second question was easy. Clapham in the '80s was possibly the drabbest and dreariest part of London – a long way from the posh and gentrified "Cla'am" it later became – but I had met and rapidly fallen in love with a pretty little secretary called Christine, who happened to work there. Throwing caution to the wind, I had left the peace and security of my parents' house and taken a small flat with her just below The Pavement.

Blonde and petite, and standing not more than five feet on her tippy toes, Christine had a body to die for. Unfortunately, she also had the brain of a 12 year old. Christine was fond of wearing the same pyjamas she'd worn since she *was* 12 years old, and of cuddling her childhood teddy bear, named Ralph. She was also fond of speaking in baby talk, as in: 'How is Mister Percy today? Does Mister Percy want to stand up and say hello?' This was addressed to my penis, which was subjected to vigorous sex at least three times daily. After a few weeks, Percy got bored of saying hello. He most definitely wanted to say goodbye. That did not stop the pea-brained nymphomaniac, however. She simply moved a new boyfriend into the spare room and started bonking him instead. 'The poor bastard,' I mentally sympathised. 'She'll wear him out in a month.' But she didn't, he apparently had a much more resilient Mister Percy than me, and they both ended up getting engaged and moving out. Which left me happily celibate and living on my own at 142 Victoria Rise for the next three years.

The answer to the first question, what I was doing working with old people, was much more complex. Yes, it did have something to do with pleasing my mother, with finding a job I could hold down for more than five minutes.

But it had a lot more to do with my grandfather.

Chapter 3

My Grandfather

I'm sure a lot of people have had remarkable grandfathers, but to me – at the young and impressionable age of 27 – mine was the most remarkable grandfather ever.

He hadn't made much of an impression on me the first time we'd met. I was only three at the time, and he had flown over from his native Hungary to console my mother, who had just lost my Polish father to a third and fatal heart attack. But if he didn't make an impression on me, I certainly made an impression on him. On his trousers, to be precise. My mother had failed to alert him to my one word of Polish – 'Kupa!' which meant 'I have to go right now!' – and since he kept dandling me on his knee instead of rushing me to the toilet, I pooped on his pants.

The second occasion I met my grandfather was even more dramatic. I was now twelve, and he was back in the UK to attend my mother's marriage to my step-father, the wonderfully named Bertram Mutton. Precocious to a fault, and unchecked by any male authority for almost a decade, I made the mistake of badmouthing my mother in his presence and instantly regretted it. A sharp slap to the back of my head had me seeing stars. 'Johnny!' admonished my grandfather, using my childhood moniker. 'If you talk like this to your mother, you must be a very bad boy!'

It was a lesson I took very much to heart.

In August of '81, a year or so before I made my rash decision to work with the elderly, my grandfather came to visit for the third time, and I was shocked. Gone was the keen-eyed and stately figure I dimly remembered from my childhood. Now he was nearly blind (from glaucoma) and badly limping from a recent operation. 'Good Lord,' I thought when he unsteadily entered the house. 'How on earth did he get on and off the plane?'

I soon learnt the answer. My grandfather had a will of steel. He also, when I sat him down one day and interviewed him (with a mind to documenting my family history) had a remarkably good command of English.

'I first learned English as I was 22 years old,' he told me.

'When I was studying at the School for Engineers in Budapest, I get lessons from an old English lady. She was the daughter of the riding master of the Hungarian Queen Elizabeth, Franz Joseph's wife. But then I get my diploma, I get a position at a great factory – making windmills! – and I have no more time and money for English lessons!'

Even more remarkable was how many times my grandfather had been up and down in his life. At the start of the first World War, he was sent as an artillery lieutenant to fight the Italians in Albania, and nearly drowned on his return. The ship carrying him home to Hungary ran itself on the rocks off the coast of Yugoslavia, and all but a few of the troops on board perished. His luck was that he had malaria and didn't jump into the freezing sea unlike most of the other soldiers. He'd felt too cold to leave his cabin, he told me, and was rescued when the ship didn't go down at all, but was salvaged.

Back from the war, he landed a very good job in an agricultural machinery factory, and did well enough to get married and have two kids (my mother and my uncle Hunor), but then the Great Depression came along and four out of five of all factory workers in Hungary were laid off. My grandfather was one of them. Undeterred, he struggled on, earning the modern day equivalent of £5 a day (making thread in an English-Hungarian factory) for the next seven years until fortune suddenly smiled upon him.

'I meet my old friend from university, Boronisza – which means "a man who does not drink wine" – at a big dinner in Budapest. He is now the Minister of Hungarian Industry, and he asks me what I am I doing. Then he invites me to come and join him as Chief Engineer for Industry, with responsibility for all Hungarian factories making steel, machinery and iron. A very big job. A beautiful job.'

'Yes,' chipped in my mother, who had just brought in two cups of cocoa. 'It was a very good position in the Ministry. His next step, but for the Second World War, would have been Secretary of State, a Cabinet position. But then his wife, my mother, died very horribly of cancer. This was in 1937, and although he held his position right through the war, he was broken hearted. It was terrible blow for him, losing my mother. He idolised her.'

My grandfather nodded sadly at the recollection. Tears welled up in his eyes.

'Yes, I lose my beautiful wife. And then, in 1945, the Russians came and I lose my position. I also lose my son Hunor. He was wounded in the fighting, and taken to Vienna, but I did not know this.'

'All that we did know,' my mother continued, 'was we had to get out of Hungary. Because everyone was bombing us. The Russians bombed us because of the occupying Germans, the Germans bombed us because the Russians were advancing in on them in Hungary, the Americans were using us as dumping ground for any spare bombs left in their planes, and even the English were bombing us – we never knew why.'

I picked up my cocoa and looked curiously at my two closest relatives. How come I had heard nothing about this before?

'Because you never asked!' snapped my mother when I posed the question. 'Honestly, John, you have as much interest in us as a family as a cat bored of its kittens!'

'Well, I'm interested now,' I said. 'What happened next?'

'What happened next,' said my mother, 'was that I nearly died. The Americans started dropping big "chain bombs" – strings of bombs chained together – and wherever they landed, a great long gash or valley appeared in the terrain. One Sunday I went to church late – I just caught the last Mass – and while I was away the Americans dropped several chain bombs and completely levelled the street where our villa had been standing. I found my father furiously digging in the ruins looking for me. He couldn't believe I had survived!'

Over the next hour or so, as I got caught up in the story and my cocoa went cold, my mother filled me in on the true horror of war, as experienced by hapless Hungarians who had lost practically everything overnight.

'Then began the task of digging all our belongings out of the rubble. This was difficult because it was winter and a thick snow covered the ground, but some army officers came over to give us a hand. We recovered quite a lot of our possessions, and loaded them onto sledges and pulled them over to the school for the university students, where the Ministry was located, and where we gained temporary lodging. I came across a school friend of mine. We had only been dancing at a ball the previous year. Now he was in uniform. And I was without a home. We looked at each other, and I said: "Well, that's life!"'

'So what did you do then?' I asked. This was the first time I had ever listened to my mother with such attention. Her endless monotone monologues usually put me to sleep.

'Well, all of a sudden, the bombing stopped and everything went quiet,' she said. 'The quiet before the storm. A few people decided to get out, and make a dash for Austria. It's strange, how in times like these, people make so much of *belongings*. You know, there were mothers and children desperately in need of transport, but the last lorries had left already, weighed down with other people's heavy luggage and possessions.'

My grandfather nodded sadly, then broke out in a big smile. 'Do you remember the swimming suits, darling?' he said.

'Oh yes, the swimming suits. By sheer luck, we managed to

get to the Austrian border. One of the Ministry officials, a friend of my father's, had a car. But by now, all that my father and I had was a rucksack and a suitcase. And because we had left in such a hurry, we had just stuffed them with the nearest things to hand – useless oddments mainly, like half-knitted skirts, and gloves, and an electric iron. I couldn't believe it when the first thing I found when I opened up my rucksack was a swimming suit!'

'And the first thing I find in my suitcase is a swimming suit also!' laughed my grandfather.

'That was so funny,' my mother grinned at me. 'We didn't even have a bowl to make pastry or collect water in when we got to my dad's friend's sawmill north of Graz, but there was a nice big lake there. We did a lot of swimming!'

I loved watching the humour and warmth running between my mother and grandfather. They had been through so much together; the bond they shared was so close. But one thing puzzled me: why had my mother gone back to Hungary from Austria when she could have come to England a year or two earlier?

My mother shrugged. 'I could have gone West, to England, at that time. But I stayed with my father, and took the first train back to Budapest with him, in May 1945, to continue the search for my brother. When we had left a few months before, we had had two places in Budapest: my grandmother's house and our own little flat. When we got back, however, there were other people living in them. That was the fortune of war. With so many people losing their homes in the bombings, they just moved into any empty house they could find. And we couldn't get them out. In fact, that was the law. Possession was ten tenths of the law in wartime. And we were treated like criminals, because we had left Hungary. It didn't matter that we had left for our safety. Anyone who had moved out at the same time as the Germans was regarded as a Nazi sympathiser. Or a revolutionary. All the Hungarians who had stayed told us we had deserted them in their "hour of need". They accused us of believing in a Nazi victory, and seemed to have expected us to just stand there and welcome the invading Russians with out-stretched arms.'

'This we could not understand,' murmured my grandfather. 'I do not hate the Russians, but they were very bad for our people.'

My mother stood back and regarded him with disbelief. 'You do not hate the Russians, dad? But they took everything from you – your homes, your lands, your job, your status – and left you with nothing! Let alone what they did to our country. I mean, we could see from the train as were coming back into Budapest, all those nice little villas and holiday places had been absolutely ruined. The Russians did absolutely everything in them and to them. They completely vandalised them. They said they were bringing "culture" to Hungary, but they brought nothing! Standards of living plummeted immediately. The Mongol troops I saw didn't even recognise a water closet when they saw it. They thought it was for drinking water from. They pulled the chain, put their cup down the bowl, and drank the water! And they used to kill you for a wristwatch. Because they didn't have watches. I've seen army officers with a string of watches – men's and ladies' watches – clear up from the wrist to the elbow. And one even had an alarm clock dangling off the buttons of his army uniform. And he didn't know he had to wind it. None of them did. So when it stopped, they just threw it away. I tell you, that was the case. And they said they brought us culture!'

There was an uncomfortable silence when my mother finished her rant. My grandfather shifted uneasily in his seat. I made a mental note never to sing 'O Chichonia' in her presence.

Then she was off again.

'So we didn't respect them. We couldn't. Mind you, we didn't fare much better under the Germans. We were very disappointed in the Germans too. They could have declared Budapest an open city. Instead of which it was left to be totally destroyed. They blew all our beautiful bridges up, just to stop the Russians crossing over from Pest into Buda, where all the fighting was. And it was winter, so the Danube froze over and the Russians just *walked* across! They didn't need to destroy our beautiful bridges at all. That was very bad of them. The Russians swept right across Hungary, and pushed up as far as Austria. And they took all the factories and dismantled them all. By the time the Americans arrived, in just a few jeeps, the Russians had been and taken everything. Even our prize racehorses. They rode them up and down and drove them so hard that we had to shoot them. They threw hand grenades into the little lake we used for swimming, and where we used to do our washing. They killed all our fish. And they did it just for fun. They rode into the lake stark naked and on horseback. And when we went to the small outside toilets in the Hungarian camp in Goss, we had to crawl on our stomachs. Because they kept shooting at us. It was a favourite sport for them. They did that for fun too.'

Yes, my mother really had it in for the Russians. In later life, when I brought home the German girl, Andrea, I wanted to marry, I remembered feeling very grateful she had not been born in Stalingrad. Okay, the Germans – despite Hungary's claims of neutrality – had invaded my mother's country in 1944 and forced her beloved brother to fight for them, but that was chicken feed compared to what she felt about the Russians.

'So how was it after the war?' I questioned my grandfather. 'And did you find your son, Hunor?'

He sighed. 'When I come back from the war – without position, without home, without anything – I was given the job of managing a factory for repairing vehicles, mainly lorries, for the Russians. The founder of this factory was a very clever man, an inventor, who invented the first engine running from benzene. I had looked after him when I was in the Ministry, now he look after me. But it was very hard work for very little money, and worse, it was great inflation in Hungary at this time, after the war. And my son is coming home wounded and ill. He had luck – he was not sent to Siberia.'

Yes, I knew about Siberia. My own Polish father had been plucked out of university and sent to an icy gulag there for four years. It was more than half the reason he had died so young.

I could see my grandfather was getting tired, but there was one last thing I wanted to know from him. Had he been present at the 1956 revolution in Hungary, when a mass of disgruntled Hungarians, beaten down by a decade of Communist repression, rose up to throw off their shackles? I thought he might be reluctant to talk about this touchy subject, the uprising had been put down with such brute force, but no, he seemed surprisingly eager to share.

'Yes, I remember the revolution,' he said with a grin. 'I was in Budapest at this time. Yet there were few televisions in Hungary, and I had no radio, so I had no knowledge that the Russians had arrived. I did not hear the news. So, in the morning after the invasion, I wake up, I eat as normal, and I am going to the factory. I am walking out of the building into Baraszda Street, and all the other people in the other rooms are calling: "Where are you going?" I said, "I am going in the factory!" They cannot believe this. "You are going in the factory? Do you not know there has been a *revolution?*' I said, "No. I had no idea." So they said "Look!"...and point. And at this moment are coming the Russian tanks, at the end of the street.'

'Wow,' I said, suddenly proud of my grandfather who could singlehandedly face up to a street load of Russian tanks. 'Weren't you afraid?'

He sniffed. 'What am I afraid of? They will shoot a single unarmed man in the street? I wave to them "Hello!" and then I am going in the factory. And it is good, because everywhere else, there is no opportunity to get food, clothing or to buy anything. In the factory, we had lorries which went out and came back with provisions and goods to eat. They brought back many chickens!'

'But you had seen this coming, hadn't you, dad?' chipped in my mother. 'You knew the Russians would come. I'd been here in England a few years by this time, and was now working as a cook for Lady Astor, and you had written me a letter, just days before the revolution, saying it was only a matter of time. It's so funny you never knew about it when it finally happened!'

My grandfather permitted himself a wry chuckle. 'The revolution could never succeed,' he said. 'We had no army. But afterwards, the Russians gave all the factories back to the Hungarian state. They had to, because the Communists were not ready to run our factories. They needed old engineers with skills like me, so I get a leading position...and five years later, when I retire at the age of 63, a very good pension. So yes, the revolution was very good to me!'

'Your grandfather is being very modest,' my mother informed me in a whisper. 'He practically rebuilt all the factories in Hungary in that short time, along German lines.'

I looked at my grandfather with new eyes. Survived shipwreck, struggled through the Depression, shrugged off two world wars, faced off Russian tanks and became a captain of industry. Wow, what a man!

'Why can't he stay with us permanently?' I asked my mother, suddenly anxious at losing the first male role model I'd had in my life.

She gave me a stern look and ushered me into the next room. 'That can't happen,' she told me in a low whisper. 'It would have been okay a few years ago, when he could see and wasn't half deaf. But his last illness – and the operation on his prostate gland – I think broke him. If anything happened to me, he would be lost. He doesn't know anybody over here in England. He would have nobody to talk to. Worst of all, he couldn't have a wireless. At least, not one with a Hungarian station. It would be worse than being in solitary confinement. And there's no way I could place him in a private home, because I couldn't afford it. He's only over here on a temporary one-year visa anyway, and I had to produce bank statements to prove I could support him financially without him becoming a burden on the state. If I ever ran out of money, he'd be sent back to Hungary immediately. His situation is very difficult. Here he can communicate, he has his radio (even though it gives him only depressingly anti-communist propaganda on the 'Free Europe' channel) and I can look after him. But his heart is back in Hungary, and so are all his friends, and this is where I feel sure he wishes to spend his last days.'

I looked at my mother, and made a silent determination.

If I couldn't help my grandfather, if I soon had to say goodbye my new battle-scarred hero, I would help others just like him.

Chapter 4

Pocket Money Day

Thursday was pocket money day, and my first chance to meet the home's residents. I wasn't looking forward to this: if Old Bill and his fencing partner Mr Keogh were anything to go by, this particular collection of old people were going to be nothing like my grandfather. To make matters worse, despite my well-practiced veneer of confidence, I was crippled by a shyness that went back to my adolescence, to a time when I was a tall, thin geek with thick National Health glasses and a stupid, pudding-bowl haircut. A target for ridicule in the classroom and in the playground, I was wary as a thief around total strangers.

It was with some apprehension that I trudged down to the post office with Mr Bragg, the home's grim-faced handyman, to collect the residents' weekly stipends. Then, ignoring Matron's snippy advice – 'Don't spend too much time with them, dear. They do talk a lot of twaddle!' – I prepared to visit them one by one and give them their money in neatly-sealed brown envelopes.

My first port of call, fortunately, was the Chairman's mother, Mrs Duff. I say 'fortunately' because she was nothing like her truculent son and seemed genuinely happy to see me.

'You must be Winnie's replacement,' she said with a big smile. 'It's so good to have some young blood in here!'

'Winnie?' I smiled back. 'Oh, you mean Mrs Gregory. Yes, I'm the raw recruit. But what's this I hear about you crying into your porridge? That's been bothering me since I first arrived.'

Mrs Duff downed the doily she had been sewing and struggled to recall the event. 'I wasn't crying *into* the porridge,' she said at last. 'I was crying *about* the porridge. I *hate* porridge!'

'Can't you get something else?'

'They won't give me anything else. Matron says it's good for my bones, but it just makes me sick. Can't you stop them keep giving it to me?'

I wondered briefly if this was Matron's way of defying Mr Parker's tyranny – taking her spite out on his poor old mum – but then dismissed it. No-one could be that cruel, surely. I also wondered who had rung in the complaint about the porridge. Someone with a genuine beef against the Chairman?

Next up was a spindly little thing called Betsy, and she had a definite naughty side to her.

'Ooh, you're a sight for sore eyes,' she informed me, grabbing both my hands. 'You're gorgeous, you are. Will you be my new toy boy?'

'Err…what does it involve?' I asked her helplessly.

Betsy gave a loud cackle. 'What does it involve? Well, if I were twenty years younger, I'd have had your trousers off by now. As it is, do you mind if I just massage your thighs?'

'Oh, don't mind her,' observed Mrs Winch in the neighbouring chair. 'She's sex mad, she is. Only way to get her off topic is to talk about the "good old days".'

'Were they that good?' I enquired curiously, detaching

Betsy's wandering hand from my right knee. 'Here, hold on a minute, I'd like to tape this, if you don't mind.'

Several pairs of eyes observed me as I leapt manically to my feet and dashed back to my office to collect my trusty Sony Walkman. Ever since I had interviewed my grandfather, it had rarely been far from my side.

'Yes, they *were* good,' recalled Betsy when I returned and repeated the question. 'Things are too fast nowadays…you can hardly breathe!'

'Everything's made for *speed* today!' chipped in Mrs Winch. 'Except Betsy!'

Betsy's hand strayed back to my thighs. 'I can move when I want to,' she said with a wink. 'Though I'm not as fast as I was…'

This time I took hold of the hand and held it. I had just noticed that Betsy was nearly blind. Touch was obviously very important to her.

'So what's your story, Betsy?' I asked her. 'You must have seen a thing or two.'

Betsy gave a hollow cough. 'I was born in 1903, not far from here. And we had to walk everywhere, 'cos there were no buses or trams or even bicycles. We walked three miles to school and three miles back…and of course the cows used to walk up the street with us. It was all fields and farms and open spaces then. Ooh, it was lovely, really.'

'We didn't have cows,' intervened Mrs Winch, 'But we did have horses!'

'Oh yes, Lady Muck,' sniffed Betsy. 'You probably had a horse of your own and rode it all the way to Buckingham Palace. Me, the only horses I saw were the ones who took the mail. When I lived in Bloomsbury – it must have been 1922 – all the Royal Mail from Mount Pleasant was taken up to

King's Cross at night, drawn by two huge dray horses. It was so dangerous at King's Cross, because there was so much horse-drawn traffic. I mean, everybody complains about all the cars nowadays, but it was much worse then. It was harder to stop a horse than a motor car! They went along at a cracking old pace, and once they'd got up steam it was a devil of a job for the driver to pull them up. I mean, my mother wouldn't let me out anywhere near Guildford Street when it was mail time. It was suicide trying to cross the road with all those horses bombing down it.'

'So,' I laughed, 'life was fast in at least one respect!'

'Ooh, did you hear that, Ethel?' said Betsy, clapping her hands together in joy. '"In at least one respect". Don't he talk posh!'

Mrs Winch, aka Ethel, nodded approvingly. 'Good education. Plus a bit of parental discipline, I bet.'

'My father was a bricklayer,' said Betsy, apropos nothing. 'There was six of us and his word was law. He never hit us or nuffink. But if he said no, he *meant* no. We couldn't have wished for a better father, really.'

'Did you ever cross him at all?' I asked.

'God! I was sent to bed. Without my tea. Though my mother used to sneak it up to me. That's the problem today, there's no discipline with children.'

'It's got to start early,' agreed Ethel. 'When I was young – I was 17 and still at school – I bought this lipstick at Woolworths. I shoved it on in the dark of the house, and my mother said "You're not going out looking like *that!*" And we argued, and she went off into the scullery and returned with something, I couldn't see what it was. And she sat down and said, "Well, I shan't let you!" And I replied, "Hmmm! How can you stop me?" And she whipped out this large jug and said, "You open that door, and you'll have this jug of water over you!" Well, I only *had* one frock, and I wasn't about to see it ruined. But I challenged her and said, "You *wouldn't!*" And she smiled and said: "You just try me, if you don't think I would!" But I *knew* that she would – even though we had no money and she'd be spoiling that good frock. And she wouldn't have thrown just a trickle of water over me. I would have got the *lot!* So I went back and removed the lipstick. And that's the difference, isn't it? Parents nowadays make a lot of threats, but don't do anything. But parents then *did*. They meant exactly what they said.'

I thought of my mother. She never hit me either, but she knew how to keep me under control. If relentless nagging in a Hungarian monotone didn't make me mend my ways, she would simply break down and cry. She knew I hated seeing her crying.

'What about schools?' I asked next. 'Were they very disciplinarian?'

'In our day,' ventured Betsy. 'You got the cane. If you did anything you shouldn't.'

I thought briefly back on my own school days, in particular on the leather covered whalebone called a ferula which did far more damage than a cane. Six strokes of that was enough to have you weeping into your sleeve the rest of the day. Twelve, which was reserved for persistent offenders like me, could split your hand open and require medical treatment.

'Or you were put behind the blackboard,' added Ethel.

'Yerse, you were *punished!*' said Betsy with more than a degree of ferocity.

'Doesn't sound so bad to me,' I said, rather puzzled. 'What went on behind the blackboard?'

Betsy's brows furrowed. 'Nothing. That was the awful

thing. You didn't know what was going on, which was very annoying.'

'I went to a "good" school,' said Ethel. 'The cane was rarely used. If you got the cane, it was a very big thing. It was the ultimate deterrent. *But* the teachers were always in charge. There was no question about the kids ruling the teachers. *Our* big punishment was, we sat with our hands on our head.'

'That's right!' piped up Betsy.

'When your hands started to flag, the teacher would say: "Elbows up!" And you try sitting for five minutes with your elbows up like that! Then she would say: "Right! You can let them down again now. Any more noise and disruption, your hands will be back! The teachers don't do that anymore…'

'They're too young,' said Betsy. 'In our day, teachers were older.'

'Rot!' snorted Ethel, leaning heavily forward in her chair. 'They were just as young. I mean, they didn't all come out of teacher training and sit around for 20 years and then start teaching, did they? I can remember one of my teachers, Miss Hunter her name was, who came into school each day on a *motorbike*. We thought she was smashing. *But* she didn't stand for any nonsense. I was very bad at maths, and I was a naughty little girl, and she used to say: "Have-you-got-it-into-your-*head?*" And she used to dig a thick marker pencil into my head with each syllable, just to make sure I *did* get it into my head!"

I looked at my watch. I knew I should be moving on – there were 30 more residents just as keen to talk to me because nobody else seemed interested – but I was warming to my task. These two lively old souls had more pluck, spirit and humour, than anyone I knew half their age. It seemed incredible to me that Matron and the other officers paid so little attention to such colourful characters – they were missing out so much on their rich life histories, which I was personally finding quite fascinating.

'What about holidays?' I ventured a last question. 'Did you have them?'

'Yerse, we had our holidays,' said Betsy. 'Not many, but we had 'em. But we never went anywhere much. I mean, all round here was country. Wherever you went, for miles, was all country.'

Ethel nodded. 'We used to go on the Sunday School "treats" as we called it, we always brought something home for our mum and our dad. We might go to the seaside, either to Brighton or to Margate. And we always brought something back. We were the first generation to do that. It was only in the 1900's that the railway started taking people places. Before that, they hardly went anywhere.'

Betsy's face had settled in a look of dreamy recollection. 'We used the tram,' she murmured happily. 'Went on a tuppenny tram ride…that's how our parents took us on outings. And you could go all round about London, you know. It was a big treat to go into London. You'd only go once in a blue moon.'

'Though many people never went out of their district,' said Ethel in conclusion. 'I know people now who have never been *out*, have never been to Central London in their lives.'

I rose out of my seat and clicked the Walkman off. That could have been me they had been talking about: I had hardly set foot in Central London except to track down rare old comics in dusty Soho book shops. As for holidays, I had inherited the travel bug from my mother – she had taken me to a different place in Europe every Summer – but then she had married old Bert and he'd taken her, me and my step-brother to the same dull place – Mevagissey in Cornwall – for as long as I could remember. Yes, I had managed few 'safe' foreign holidays since then – to Rome, Athens and Israel – but part of me craved something far more free and adventurous, where I could live out the heroes of my childhood: as a dashing rogue of the sea from one of Rafael Sabatini's novels, perhaps, or as an 'into the wild' explorer in the style of Jack London. Many were the times I would wistfully look through travel brochures and think of far off places with exotic sounding names like Bangkok or Kathmandu, but doubted whether I would ever have the money or the courage to go there.

'Thank you ladies,' I said as I prepared to move on. 'That was very interesting. I'll be back for more on another occasion, if you don't mind?'

'Ooh, he's got manners, don't he?' said Betsy, snapping out of her trance.

'Yes, indeed,' agreed Ethel. 'You can't buy good manners.'

*

About two hours later, I came to the end of my rounds and stared at the single remaining brown envelope in my hands. It belonged to Old Bill.

'This is not like Bill,' I thought to myself. 'He kicked up such a fuss about his money when he thought I was the Mayor of Lambeth. Why isn't he collecting it now that I actually have some?'

I eventually found my lost sheep upstairs in his room, lying on his back and looking extremely uncomfortable.

'What's the matter with you, Bill?' I enquired softly. 'Are you in pain?'

Bill gave a brief nod. 'It's my leg. I keep telling 'em it hurts but nobody believes me.'

Too Young to be Old

I was about to comment further when I took a proper look at the leg. I hadn't noticed it before, because I had never seen Bill horizontal, but it was at such a weird angle to the body that it was obvious something serious was wrong.

'Here,' I summoned Miss McCann to my side, 'come and have a look at Bill's leg.'

'What about it?' said the ever-stressed deputy Matron. 'Honestly, Bill, what trouble you causing now?'

But then she noticed the weird angle of the leg and went pale.

'So you see it too?' I asked her. 'It doesn't look right, does it?'

'No, it doesn't,' she said in a flustered whisper. 'Can we talk about this outside?'

Putting Bill's money gently on his nightstand, I followed Miss McCann out into the corridor and listened to a truly awful confession.

'Please don't tell anyone about this,' she pleaded. 'We could all lose our jobs. Bill fell one day, and afterwards had this real bad pain in his leg. But – largely because he'd cried wolf so many times before – the home's doctor took one look at him and said there was nothing wrong. So we pushed him to keep exercising, but he kept going on about it hurting. So we took him to hospital and he was kept in casualty, still being exercised. And then he came back, still complaining, and we thought he was still putting it on, so we pushed him further. He went back and forth between his GP and the hospital four times before we found out what was really wrong. A specialist finally got hold of him, and told us that he had a shattered hip! He'd had it since the day of the fall. The guy who looked at him saw it immediately. He was appalled. Especially when we told him Bill had been walking round on this shattered hip for three months. He put a metal bolt into his leg, but all the bone had long since fragmented into the muscle, and he could never walk properly again. It was a case of pure negligence.'

There was a pause as my jaw dropped, and then a voice behind us said: 'I heard all that, Miss McCann. You're fired!'

The pale deputy spun around in confusion. And was confronted by a red-faced Mr Parker, who had just emerged from his mother's room.

'Oh, oh, but what did I do?' she protested. 'We all knew there was something wrong, but we all missed it!'

'What did you do?' echoed the Chairman nastily. 'What you did was cover the whole thing up. One sniff of this, and the Social would shut us down without farting! Now, you get yourself downstairs and tell everyone who knew about this – and I mean everyone – to eff off out of my building!'

'Erm...is that a good idea?' I intervened feebly. 'Won't that draw a lot of unwanted attention to the incident?'

A pair of angry little piggy eyes stared at me for what seemed an eternity. Then the penny dropped, and they gave a single blink.

'There you go again, Mr Kusy. You're absolutely right. Can't have the Social witness a wholesale sacking without them asking lots of awkward questions. Brilliant. Okay, then, nobody gets sacked. But somebody's got to bring the whole miserable lot of them into line. Now, who do you think that could be?'

Chapter 5

In the Hot Seat

They say be careful what you wish for, and so it was with me. I should have been happy when Mr Parker assembled his committee and appointed me – after three short weeks – the home's new Director, but I wasn't. I had spent my entire early life – starting with school teachers and then the Jesuits – being a rebellious challenge to authority. I had no experience whatsoever in wielding it. Manage a home of 36 residents and 16 staff? I could barely manage my shoelaces.

What I should have said, looking back on it, was: 'Right, you lot, the party's over. No more crafty fags in the toilets, no more suspect "sickies" and definitely no more treating old people like troublesome children. There's a new kid in town and his word shall be Law!'

What I actually said, when I nervously convened a staff meeting and saw all the bored, resentful faces staring back at me, was: 'Erm, I'm as surprised by this appointment as you are, but don't worry, I want us all to get on. I want us all to be friends.'

There was a stunned silence as everyone digested this limp-handed display of power, and then a deep, resounding guffaw echoed around the bare walls of my office.

'Friends, is it?' mocked Mr Bragg, his lips curled in a sarcastic sneer. 'Okay, we'll be your friends, laddy – if you get us all contracts of employment. Yes, you stop Parker sacking us left, right and centre, and we'll be your friends alright!'

I fought the urge to run out of the building in tears. What right did this jumped-up handyman have to talk to me like this? And what was with the "laddy"? Everyone knew I was the youngest person in the building. Did he really have to bring it to their attention?

'You don't have contracts of employments?' I stuttered, trying and failing to keep my voice steady. 'That's news to me.'

'Oh, the senior staff have them, dear,' chipped in Matron. 'Just not the care staff. Oh, and Mr Bragg, of course.'

'Neither do we have terms and conditions of employment,' added the irascible Scotsman. 'Which means we can be slung out on our ear whenever Parker or any of the management committee feels like it.'

I shifted uncomfortably in my chair. Mr Bragg had just handed me a Gordian knot of a problem. How on earth was I going to cut it?

'Err, I'll do what I can,' I assured him as I closed down the meeting. 'Yes, leave it with me and I'll get back to you soon, okay?'

The boiler-suited handyman's protracted look of disdain told me what he thought of that offer, but it was all I had to give.

There was no way I was going head to head with Mr Parker about those contracts of employment. What was there in it for him? Absolutely nothing. I was going to have to get somebody else to do my dirty work for me. It was an old tactic, one I had perfected at Jesuit school. I still remembered the two skin headed thugs I'd engaged to scare off bully boys in the playground. Their names were Kev and Trev, and I'd done their homework for them for five long years, until they got expelled for flushing a holy crucifix down the toilet.

I chanted a long time before a name came to me – Mrs Teasdale. She had been kind to me when I had first interviewed for work at the home, and she had been the only member of the Committee who had backed Parker's bid to make me Director at such short notice. All the rest – especially flush cheeked Mr French, the Treasurer – had doubted my ability to handle such a responsible job with so little experience or training.

'Ah yes, Mrs Teasdale,' I thought excitedly. 'She's the one for me. Parker will never see her coming!'

But Mrs Teasdale, when I convened a private meeting with her the next day, was doubtful.

'Do you realise the implications of what you're asking?' she said. 'If we give all the care staff contracts and terms and conditions of employment, we would be obliged to give them paid holidays and sick leave. Plus they could do us for unfair dismissal. I can't see Parker buying that – he has unfair dismissal down to a fine art.'

I surveyed the prim, bun-haired figure before me, and took a stab in the dark.

'What do you know about Mrs Duff's porridge?' I said.

'Mrs Duff's porridge?' she replied, looking suddenly flustered. 'What do you mean?'

Her eyes told me everything. The left one in particular, which was twitching like a rabbit's nose.

'Well, I couldn't help noticing,' I said innocently. 'But Mrs Duff's files state quite clearly that she hates porridge, yet she is being given it every day. It was you who supervised her admission to the home last month, wasn't it – didn't you notice that too?'

Mrs Teasdale's lips pursed in a tight grimace.

'I can see where this is going, young man. Yes, I did notice that. And yes, it was me who told Matron that Mrs Duff loved porridge, no matter what she might say to the contrary. It was also me who phoned in about Mrs Duff crying into her porridge. The Chairman didn't know it was me, I disguised my voice, but it was worth it just to see him squirm for a change.'

I stared at the discomfited deputy Chairman in amazement. This was beginning to bear all the hallmarks of a TV soap opera.

Mrs Teasdale sighed, and sat down. 'You may think me cruel, using a poor old woman as a means of shaking Parker out of his comfort zone, but she was a wicked old hag before she lost her marbles. The home was my idea. I should have been Chairman. But she stole it from me and got her son appointed instead. Since then, I've had eight years in his shadow, watching my lovely home for the needy elderly reduced to a circus sideshow by his bully boy scare tactics. I'm ready to scream!'

Yes, 'circus sideshow' just about summed it up. I sat down opposite her, and rang for a pot of tea.

'So what do we do now?' I asked her. 'It seems to me that the home is being run by the care staff, with Matron and all the officers running around like headless chickens. How am I supposed to please everybody and not have a full scale revolt on my hands?'

'I'm not oblivious to your situation,' said Mrs Teasdale, regaining her composure. 'You're caught between a rock and a hard place, aren't you? Leave those contracts of employment to me. It won't be easy, but I'll find a way of talking Mr Parker round. Oh, and I'm not doing it because I fear what you might say about the porridge thing, I really don't care what happens there. I'm doing it because it is the right thing to do. The staff *should* feel safe in their jobs!'

*

Later on that day, leaving that little pot to come to the boil, I heated up another.

'I have no idea how to manage people,' I was forced to acknowledge. 'I'll have to get someone to do it for me.'

The someone I had in mind was a calm, unflappable care officer by the name of John Gray. Tall, balding and possessed of a huge brush moustache, he was the only member of staff who seemed unafraid of Matron. Indeed, when she went off on one of her rants, he would just stand there and survey her with a mild air of amusement.

I found John at the back of the large wood-panelled library which opened out to the garden. He was in consultation with a Mr Bartlett who had come for his daily discussion about his bowels.

'Nothing's moving,' Mr Bartlett was saying. 'I've been in the loo for an hour past, and nothing's moved.'

John winked at me, and indicated I should sit down.

'You're still on the Dorbanex?' he told Mr Bartlett. 'And it isn't working? Well, let's try you on Lactulose.'

Mr Bartlett's face darkened with suspicion. 'What's that?'

'Well, it does the same sort of thing, but it's a clear liquid instead of orange. It does the same sort of thing. They all do exactly the same sort of thing. You can try it if you want.'

'Why isn't it orange?' said a persistent Mr Bartlett.

'Because it hasn't got a dye in it like Dorbanex. Some laxatives are orange. And some aren't. This one isn't.'

'Well...I really don't know,' sighed Mr Bartlett. 'I've got used to the orange one.'

John reassured his constipated guest that 'clear' was the 'new orange' and then, having seen him safely out of sight, he turned to me and said: 'You've got something on your mind. What is it?'

I didn't see any point in beating around the bush. Taking a quick look around to make sure we were quite alone, I said: 'How do you fancy helping me out, John? As in...erm...being my deputy?'

John's slim, white fingers raked his moustache as he considered my proposal.

'Hmm...well, that would depend. Would I have total authority over the officers and staff...including you know who? Because down to the inefficiency of Matron and divisions between the officers, the care staff have been used to running the home for years.'

I nodded. 'You saw how it went with my meeting yesterday. I'm in over my head with both Matron and Mr Bragg. What I need is someone I trust to keep them in check while I crack on with important business...like getting everyone their bloomin' contracts of employment.'

'What you need is a miracle,' said John. 'But okay, I'll be your huckleberry. Can't promise you Mr Bragg, he's a real hard bastard. But Matron, yes, piece of cake...'

*

Outside, it was snowing. I looked down at my thin shoes, which would soon not be suitable for crossing the wide, frozen expanse that was Clapham Common. No, I thought grimly, it would have to be boots tomorrow, and a long leather coat.

'Here!' A thin, frail voice shook me out of my reverie. 'Come over here!'

'Oh, hello Betsy,' I greeted my frisky new friend. 'What can I do for you?'

'I got to tell you this,' said Betsy in an urgent tone. 'I had a dream about you last night.'

'Oh, did you?' I responded cautiously. 'What were we doing in it, playing Scrabble?'

'Not bloody likely. We were doing a lot more than that. We were up to all sorts!'

For one fleeting second I found my mind entering Betsy's dream world. Then it fled screaming.

'I haven't lost it, you know,' she gave me a suggestive wink. 'I still get "feelings". If I were 30 years younger, I'd give you a run for your money!'

Then, without warning, she whipped out her new toy – a small plastic hand-fan – and directed it at the crotch of my trousers.

'A young man like you must get hot in this place,' she informed me meaningfully. 'Let me cool you off a little.'

I blushed to my roots, and desperately tried to think of a way to flee my 80 year old groupie.

'I'm sorry, Betsy,' I said, 'but I have to go home now. I've really been through the wars today.'

'Been through the wars, have you?' said a second voice behind me. '*We* had a war. We had it right *here!*'

I turned around to find Betsy's pal, Ethel, shuffling up to join us. Dressed in a wide flower print dress and smoking a flavoursome Gitane cigarette, she looked set for a long conversation.

'Which war was that, then?' I asked her, looking at my

watch and ruling out an early night. 'The first or the second?'

'Oh, the first war wasn't nearly so bad,' coughed Ethel as she eased her large frame into a chair. 'They were only Zeppelins that came. But the second one, we didn't know what a good night's sleep *was* – not with all those "doodlebuggers" flying over! We slept out in the garden, in the shelter. The shelter was buried in the ground. It had a corrugated iron roof. We'd be out there, shaking like leaves, even in the freezing cold in the dead of winter.'

'Terrible, it was,' agreed Betsy, lowering the hand fan and giving my windswept assets a much-needed respite. 'Penge alone had one hundred flying bombs drop. In just one square mile of Penge. And all this in just a three month or so period.'

I fished out my Sony Walkman again. This was getting interesting.

'I will say this about the war,' said Ethel, a serious look on her face. 'It did bring people together.'

'Because they were all scared to death,' said Betsy.

'They all shared things. Not like now.'

Betsy nodded. 'But in all the areas in England which *didn't* have bombs, they couldn't have given a damn about each other. I had friends in Leeds. They were most annoyed up there at having to have evacuees from London – even those who were *rolling* in money up there, working in the woollen mills. My friend Iris went up to Leeds and showed a shop her soap coupon, and they just laughed. They hardly knew what rationing was.'

'They didn't know there was a war on!' chipped in Ethel.

'Yes. Outside the bombed areas of England, there was none of this "comradeship" everyone goes on about now. They couldn't have cared less. It was the Americans and the Canadians who sent food parcels and clothing to London. They sent far more than our own people in say, Leeds!'

Ethel stabbed out one cigarette, and lit another.

'I went up there during the war,' she ranted. 'To Leeds. And there was two women in a bus queue complaining: "We don't want those London kids up here – we didn't ask for a war!" And I said: "Well, neither did *they!*" And they said: "Well, we had a bomb dropped near *Leeds!*" They'd had a *bomb!* So they didn't want to know about London's problems. Even though this one single bomb had been dropped on them by accident, on the outskirts of the town. They had no idea how bad things were in London. And they didn't want to know.'

What I didn't know, looking back on it, was that if things were bad in wartime London, they were about to get much worse in one small corner of latter day Clapham.

A small, angry six stone bombshell was about to land.

Chapter 6

I knew Elizabeth Taylor

The following week, as we approached Christmas, I welcomed Miss Margaret Pratt into the home. Though welcomed was perhaps not the right word. Dressed in a smelly old duffle coat and wearing a glare that would have turned Medusa to stone, 'Maggie' Pratt promised to be even more of a handful than Old Bill.

'Where am I? said Maggie.

'In a very nice home for sophisticated ladies and gentlemen,' I replied soothingly. 'You'll like it here.'

'Well, that rools me out,' she spat. 'I'm about as sophisticated as a coal bucket!'

'Now, now, it can't be that bad. You can always have visitors. Don't you have family?'

'No.'

'Friends?'

'No.'

'Pets? We can always arrange for you to see them.'

Maggie shot me a look of pure venom.

'I fucking *hate* animals!'

I rolled my eyes and sighed. If this was the Committee's

idea of filling up vacant beds in a competitive market, I didn't like it.

'You sit here, Maggie,' I said, showing her to the dining room. 'Lunch will be up in a minute...'

'Oo's *this* woman?' snarled Maggie, eyeing one of her new co-diners with suspicion. '*I* don't know 'er!'

'She's a very nice lady, just like you. Her name is Miss Sherring. She used to be a Tory council member.'

'I don't like 'er. She looks stuck up, she does. Look at the gob on 'er – she looks like she just swallowed a lemon!'

'Well, *try* and like her. Say something pleasant!'

Maggie thought for a moment, and then leant forward conspiratorially.

'You won't believe this,' she confided, 'but I used to know Elizabeth Taylor. Oh yerse, me and Liz go *way* back. I used to char for 'er when she woz in London shooting that film, the Vips!'

'The "Vips"?' sniffed Miss Sherring haughtily. 'Don't you mean, the "V.I.Ps"– Very Important People?'

'I know wot I mean. And "Vips" it woz. Liz said it stood for "Very Inebriated Prat," 'cos Burton was on the piss again and couldn't get it up in the bedroom department.'

'What's she talking about, Mr Kusy?' complained Miss Sherring. 'She's making all this up!'

'That's as may be,' I said, panic rising in my chest. 'But look – here's your lunch. Chicken supreme, you'll like that.'

'*Fuck* chicken supreme,' interrupted Maggie. 'It smells of peppers. I *hate* peppers!'

'Well, just pick them out...I'll do it for you, if you like.'

'Liz hated peppers too. Gave her gas.'

'Can I sit somewhere else, Mr Kusy?' Miss Sherring demanded. 'This woman is quite deranged!'

Maggie's lips curled in an ugly sneer. 'Don't like the company, do yer? Well, I'm telling you, you toffee-nosed bitch, me and Liz were like two peas in a pod. I told her one day, I did. I said: "*'Ow* many husbands have you 'ad now, Liz? Six? Seven? I dunno why you bovvered. One cock's enough for *me!*"'

Miss Sherring moaned in dismay, and fell back in her chair.

'Okay, Maggie, I think it's time we left,' I said, rapidly reconsidering my career in care management. 'I have a nice little room where you can eat all on your own.'

*

I returned to the dining room to find Matron newly arrived and in a high old dudgeon.

'I just heard what happened!' she protested, her pink rouged cheeks puffed with annoyance. 'The staff may be yours, dear, but the residents are *my* responsibility. You should have left Miss Pratt to me!'

'We did try,' I replied. 'We rang all around the building, but we couldn't find you.'

'Well, I had my hair in rollers,' Matron said defensively. 'I can't be *everywhere* at once!'

All around us was confusion. The high walls of the wood-panelled dining room echoed with the scraping of chairs as some residents left in disgust at Maggie's antics, while others clustered around Miss Sherring and tried to console her in her affronted outrage.

In the middle of all this, Mr Parker burst in and announced to nobody in particular: 'I've just had a steaming row with the Social. I'm not in a very good mood. My wife's called three times, and wants to know why I wasn't home two hours ago.

Too Young to be Old

I've just about had enough!'

Everybody, even Miss Sherring, froze. Whatever was this bushy browed madman going to say or do next?

'Oh, excuse me,' he said, darting suddenly to the left. 'Can I have a chat with you in the office?'

The object of his attentions was Mrs Hyde, who was the daughter of little Betsy. Cool and frosty, she was as unlike her saucy mother as it was possible to be.

'I'm not here very long,' said Mrs Hyde evasively. 'Yes, I should be getting along.'

Mr Parker erupted. 'Well, alright – you please yourself. If you're not interested in trying to sort this problem out, Mrs Hyde, well, that's it! You know, if you can't spare a couple of minutes...I'm not bothered!'

Mrs Hyde fluttered her eyelashes and looked injured. 'Mr Parker! I don't know *what* you're so upset about!'

'I'm going home. Now.'

'I was just going to get my bag. Then I can see you.'

'Oh, well,' grumbled Mr Parker. 'I apologise.'

'I was just waiting for a wheelchair to take my mother somewhere.'

'I apologise...I apologise!'

Mrs Hyde's voice built up to a shrill whine. 'I wasn't trying to be awkward...'

'I APOLOGISE!' said the apoplectic Chairman. He was so dizzy with rage that he had to grab hold of a chair to stop falling over.

I watched as both he and Mrs Hyde filtered off in the direction of my office, and then I turned to John Gray, who had just come on the scene.

'What was that all about?' I quizzed him. 'What's the problem with Mrs Hyde?'

'Her mum has run out of money,' said John. 'All the equity from the sale of her house has gone on paying her fees here.'

'What?'

'It's not our problem at all, but the Social won't pay – they only pay a flat rate of £110 per resident – and our fees are £124, so Betsy is running £14 short on fees per week. And the Social won't take responsibility for her. So we've had to give her notice to quit. That's why the Chairman's in such a stink – imagine the bad press we'll get if we're seen to kick her out into the street! Mrs Hyde would have a field day with that. Our name would be mud. But what's the alternative? A few more like her, defaulting on fees, and we'll not be making enough money to pay the staff's wages.'

I stared at John. 'How come I didn't know about this? Did it all happen over the weekend? And why can't Mrs Hyde pay the excess? That's only £14 a week.'

'Yes, it all kicked off Saturday morning, when you were off duty. And that's the line we're going to have to take. Mrs Hyde can afford it. The alternative, of course, is for poor, frail Betsy to go back and live with her. Or go into a cheaper home, and there's not many of them nowadays. The idea of putting a borough ceiling on Social help, presently £110 a week per resident, is aimed at doing away with private homes who have been exploiting the market. But it doesn't take into account small voluntary homes like ours. We're caught in the middle.'

This rang a bell. I remembered the stormy Committee meeting of a week or two before, with Mrs Teasdale raging: 'I don't want to admit infirm and mentally disabled people! We'll become nothing more or less than a nursing home!' And the Chairman shooting her down with: 'We got no choice. We're in danger of being closed. That's all we're being offered. The Social and the Council won't subsidise people with some mobility and wits about them. They're only going to give us wheelchair cases and basket cases from now on. Chew on that!'

And they had chewed on that, and the result was...Maggie Pratt.

*

The scale of the problem presented by Maggie became evident two days later, when John Gray escorted her to a big London hospital for an appointment.

'It was a nightmare,' he reported back. 'The moment we arrived at Leicester Square tube (**underground railway*) station, we hit a problem. Maggie was terrified of the escalators. And she refused to go up them. It was about 9am on a Monday morning – just about the worst time we could have arrived – and she was deluged by a tidal wave of rush hour traffic. She just sat down on the floor at the bottom of the escalator and let everybody try and climb over her. "Help! Why don't one of you bastards help me?" she was screaming as she grabbed at people's ankles and trouser legs. "I'll help you! I'll help you!" I said. "Don't panic!" And all she could offer was to *sit* on the escalator while it was going up. I told her this was impossible – if she got her dress caught, she'd have had it. At this point, a whole troop of school kids turned up and practically stomped Maggie into the ground on their hazard ascent. So I thought, "Right, that's it!" and went and got the escalator stopped. On a Monday morning. At Leicester Square. In the middle of the rush hour. Can you imagine it?"'

No, I couldn't imagine it. It sounded like the stuff of nightmares.

'So,' he continued, 'while this swaying mass of people accumulated behind us, I slowly took Maggie up the escalator by foot. It took twenty minutes. I don't know if you've been on the escalator at Leicester Square? Well, it's *massive*. I smoked four cigarettes going up it – *four!* And of course she was gripping the handrails on both sides with grim determination, and nobody could squeeze past her. It was bedlam.'

I nodded sympathy. 'But you got her to the hospital in the end?'

'Oh yes, she made her appointment, but then I decided to take her for a meal. At a fish and chips place. And halfway through our meal I suddenly felt the urge to check my jacket. It felt somehow "light". And so it was, because my wallet had gone. Somebody had pinched it. So there we were where, halfway through two meals we couldn't pay for, with me going frantic worrying how I was going to explain it to the waiter, and Maggie howling: "Oh! He's has his wallet pinched! Silly bugger can't pay for the meal!' all over the restaurant. Well, I eventually persuaded the guy to take a cheque and set off back home, groaning inside at the prospect of having to go down Leicester Square tube station again. I had to get the escalator stopped once more, to take her down it. It was *dreadful!* And she moaned the whole way back on the train, telling all the other passengers: "He's had his wallet pinched! We were sitting in the fish bar, and he had his wallet pinched! We couldn't pay. No, we couldn't!" And I was trying to tell her to shut up or I'd throttle her, and then she got it into her head that I was trying to attack her. "He's a devil, this one!" she started shouting. "He's going to beat me!" And I could see all these old biddies sitting around waggling their heads. "Ooh!" they were going. "How shocking!" I tried to calm her down, and gently took her arm, but she wrested it away and flung herself to the floor. "Don't *do* that!' she began screaming. "Don't touch me!

Too Young to be Old

Ooh, he's a devil!" And all these women rushed over and shielded her from me, muttering: "Don't worry, dear...*we'll* protect you!" God, she's a nasty piece of work.'

*

While I was mulling over what to do with Maggie – and it was a long, hard mull because I could see absolutely nothing I could do with Maggie – I came across her file.

'Sent into care following a string of suicide threats,' someone had written, 'brought on by paranoic fear of the front door buzzer. Each time it went, she was sure there would be a "black man" on the other side. Never washes, smells a lot, wears the same duffle coat that she's worn for the past 30 years, plus bright purple drainpipe slacks with stirrups under the feet. Refuses to submit this clothing to the home's staff for washing. She is at a total loss to understand why people stare at her all the time. Was sent in by social worker after last suicide threat, when she phoned up a neighbour to announce that she was about to bump herself off – in full knowledge that the neighbour was laid up in bed with a broken back. The neighbour was just restrained from crawling out of her bed to Maggie's aid. Maggie promptly admitted to the home.'

'Oh dear,' I silently panicked. 'This is even worse than I thought. Whatever *am* I going to do?'

*

Just then, the phone rang and it was Brenda.

'Your month is up,' coughed my old editor friend through one of her liquorice flavoured hand rolled fags. 'Did you get what you chanted for?'

'Oh yes, indeed,' I replied. 'Now I'm Director of the home and everybody hates me. Plus I got the most difficult old lady in the world on my hands.'

Brenda laughed. A deep, throaty laugh which belched forth as a single explosive gunshot.

'I think it's time you went to a meeting,' she said.

'A meeting? With other Buddhists?'

'Well, you can continue to practice on your own,' advised Brenda. 'But you might get a quicker solution to your problem if you share it. Besides, discussion meetings are the cornerstone of our Buddhism, they're often great fun.'

So I went to a meeting and it was nothing like I had expected. For one thing, it wasn't held in a temple or a monastery, but in a large, whitewashed building called The Richmond Centre in south-east London. Second, despite the poster outside saying: 'Come to Nichiren's meeting!' there was no evidence of the bald old person pictured on it being present. 'Where is he?' I wondered to myself. 'Is he going to shuffle in once we've all taken our seats? Or is he going to spring out of that big, black lacquered box everyone seems to be chanting to?'

The answer was neither. 'Nichiren is the 13th century Japanese priest whose teachings we follow,' Brenda informed me when I voiced my concerns. 'He's not going to be shuffling in or springing out of anywhere. Oh, and that box is called a *butsudan*. It houses the biggest *gohonzon* – or supreme object of worship – in the U.K.'

Supreme object of worship? I didn't like the sound of that. It brought back memories of poor Jesus on the cross and the Jesuits who made me pray to Him.

But Brenda was ahead of me. 'Don't worry,' she said as the doors to the box were slowly opened. 'It's not a "god" or anything "outside yourself". It's just a devotional scroll, an aid to concentration. Nichiren inscribed the original one, of which this is a copy, shortly before he died. It depicts the human life in the Buddha state. Chanting to it helps bring out the Buddha state in you and in all of us.'

I gazed at the long, white scroll with strange black characters on it. This Nichiren certainly had artistic talent. It looked beautiful.

All of a sudden, a bell sounded and everyone whipped out little liturgy books and went into a long, urgent praying session called *gongyo*. This went on for about twenty minutes and then they launched into ten minutes of even more urgent chanting of Nam myoho renge kyo. The walls of the brightly lit room literally vibrated with the joy and enthusiasm of their efforts, and I found the hairs on the back of my head standing up. 'Wow,' I whispered over to Brenda. 'What a buzz!'

The first thing I noticed when the chanting stopped and I looked around the room was the incredible diversity of the people present. Not only were there all sorts of nationalities – African, Indian, even Malay and Thai – but all sorts of backgrounds too. I saw a city gent in a bowler sitting next to what appeared to be a bag lady, and a punk with a Mohican in deep conversation with a prim and proper mother of two.

'There's no distinction between people in our Buddhism,' Brenda informed me. 'And before you ask, there's no inequality between men and women either. We all achieve Buddhahood "as we are" – no need for funny clothing or shaved heads, though that punk guy might not agree with you!'

To commence the meeting, a small, dapper Japanese man stood up and introduced himself. 'Hello,' he said, 'my name is Kazuo Fuji, and I'd like to pass on some guidance from our mentor in Japan, Daisaku Ikeda. This is especially important for any of you here who are new to the practice and do not understand what our Buddhism is about.'

I felt my eyes drooping. The combination of the warm, stuffy room and the prospect of being lectured to reminded me of Jesuit school and made me suddenly very tired.

'Wake up, Frank!' Brenda nudged me urgently. 'This is important!'

And so it was. Kazuo drew forth a slim volume from his shiny black attaché case and what he read out really got my attention:

'The human being is not a frail wretch at the mercy of fate. Shakyamuni, the original Buddha, insisted that to change oneself now is to change the future on a vast scale. The Western impression that Buddhism is all about meditation is alien to the spirit of Shakyamuni. The goal of Nichiren Buddhism is neither escape from reality nor passive acceptance. It is to live strongly, proactively, in such a way as to refine one's own life and reform society through a constant exchange between the outside world and the individual's inner world.'

'Isn't that great?' said Kazuo, closing the book. 'Buddhism is action. It is not about hiding away from the world. And each one of us can achieve Buddhahood right now, by changing ourselves, by doing our own human revolution, by contributing to society. It is the only sure way to world peace. Now, does anybody here have an experience?'

Brenda gave me an even more urgent nudge. 'Go on, then!' she hissed.

I stood up, and then sat down again. Then raised my hand.

'I'm sorry,' I said. 'I'm new to all this. This is my first meeting.'

Cries of 'Welcome!' and 'Well done, you!' reverberated round the room.

Too Young to be Old

I stood up again. 'Okay, well, thank you. I'm not sure if this counts as an experience, but I've been chanting for a month and I got exactly what I chanted for – to be put in charge of an old people's home – but I'm not sure I want this responsibility. I've got a volcano of a boss, none of my staff trust or respect me, and one of my old ladies is a borderline sociopath.'

'Congratulations!' shouted everybody. 'That's fantastic!'

Brenda looked up with a grin. 'Don't expect sympathy here. These people just thrive off difficulties.'

'Difficulties are great!' the plump, spiky-haired woman next to her told me. 'Struggling against and overcoming difficulties enables us to develop tremendously. That's why we're so happy for you – some of us chant *years* to get such an opportunity to grow!'

'Yes,' agreed Kazuo, 'but don't let us frighten you with this concept, it's not an easy one to grasp at first. The main thing to remember is that all of us have Buddhahood inherent in our lives. Even a hardened criminal, if he snatches a child from the jaws of a speeding car, is, for that moment at least, a Buddha.'

I bridled. 'Does the same thing apply to vindictive, foul-mouthed old ladies with all the charm of an anaconda?'

'Yes, even to them,' smiled Kazuo. 'Buddhism is compassion, and the law of cause and effect is strict. My advice to you would be to chant for her happiness and the happiness of everyone in your work environment. Then, without fail, you will become happy yourself.'

Chant for Maggie Pratt's happiness? Was he joking? The only thing that would make Maggie happy would be for everyone else to be miserable.

But then I went back to Clapham and took Kazuo's advice, and lo and behold, not two days later, a long lost relative materialised out of nowhere and took Maggie home to live with her.

'Problem over,' I thought happily. 'Now we can get back to normal.'

Yeah, right.

Chapter 7

A Contract is a Contract

The day the staff got their contracts of employment, I nearly lost my job.

Monday, the 16th of December, started out like any other. I entered the home, unlocked the post box, unlocked the office door, unlocked the telephone dialling lock, unlocked the filing cabinet, and commenced opening the mail.

Minutes later, Matron, who had just come on duty, turned up for her constitutional half-hour chat. In between showing me the progress of her new ballroom gown, choosing the new wallpaper designs for her room out of the just arrived catalogues, she complained that a) the admin memos sent to her from my office were an insult to her intelligence; b) the fact that an inch of paper cut off her new rota for assistant staff, now pinned to my board, smacked of a nefarious plot to undermine her authority; c) could she be lent £2 for bingo tonight?

Yes, Matron and I were getting along at last. But only because John Gray had not so much undermined her authority, but quietly usurped it. All the time she spent in my office or dolling herself up in her room or chatting inconsequentially to all and sundry, John was getting on with the actual business of running the home: settling staff disputes, mediating with awkward suppliers, making sure the residents were well attended to, and generally making himself useful.

John was also making daily reports to me, so that I could give the Committee up to date information on what needed doing and undoing. Though this particular day, he had a far more pressing issue, one that I could totally relate to.

'You just wouldn't credit this!' he said as he stumbled into my office. 'I've just been chatted up by a 96-year old man!'

'Get away,' I said. 'Who was it?'

'Charlie Godbolt. I was getting him undressed for his bath

this morning, and I was helping him out of his pyjama bottoms. So he was holding onto my arm while I doing this. "It's alright for you," he says. "You're young, and you're a good-looking bloke." "Oh, thanks," says I, and – feeling it only fair to return the compliment – return with: "I bet you were good-looking once, when you were young." So he looks me up and down, and comes out with: "Oh, you've got a lovely, firm young body." I didn't know where to look. I just thought: "Oh God, I don't believe this!" Then he says, "Yes, I bet under all those clothes, you've got a really nice muscular body!" So I said, "Yes, it's all right!" and wondered whether to give him a cold shower instead of a hot bath.'

I laughed, and started to commiserate with my distressed deputy. 'I've got one of those,' I told him. 'Her name is Betsy, and she's not just touchy feely. She also has buzzing appliances...'

Just then, there was a sharp knock at the door and Mrs Teasdale walked in. Her cheeks were flushed and the bun in her hair was wobbling with excitement.

'I've got to have a private word with you, Mr Kusy,' she said. 'If you don't mind, Mr Grey...'

The pinch-faced Deputy Chairman waited until John had left the office, and then turned to me with a look of triumph.

'Well, here they are,' she said, brandishing a thick folio of printed papers. 'Your contracts of employment, and terms and conditions of employment too. As of right now, every single member of staff will have them!'

'My word,' I stuttered, taking the pile from her and fingering through it in disbelief. 'How on earth did you accomplish that?'

Mrs Teasdale's thin lips widened in what could only be described as a cheeky grin. 'I reminded Mr Parker of what went on in the cloakroom at the Conservative Party Conference in 1970.'

'You and him?' I said in shock.

'Yes, but keep that *strictly* under your hat. We were both drunk at the time, and we were both excited to be planning the home. It was before Mrs Duff stuck her oar in and ruined things. Oh, but you should have seen the look on Mr Parker's face when I jogged his memory about it. He did *not* want that getting back to Mrs Parker.'

'So he's agreeable about the contracts?'

'Agreeable is not the word I would use. "This is outright buggering bribery!" he said, and walked off in a right huff. I haven't seen him in days.'

The pleasure I got in handing out those contracts of employment to the staff was indescribable. It was the first real sign that I could do the job for which my young years and inexperience had apparently made me so unsuitable. Matron was gobsmacked, and so was Mr Bragg. 'There you go,' I felt like saying to the snarky handyman as I gave him his papers. 'Stick that in your mealy mouth and suck on it!' But then I remembered that even he had Buddhahood and contented myself with a short victorious nod. As for the rest of the staff, they instantly stopped all their dark gossiping and talking behind my back, and were – for that day at least – all smiles and laughter.

The only fly in the ointment, of course, was Mr Parker. I had also not seen him in days. How was the irascible Chairman going to react to all this?

I did not have long to find out. No sooner had I returned from my pleasurable tour of duty – in which I felt like a king distributing honours to his poorly-treated and grateful supporters – I was confronted by the terrible tyrant himself.

'I just been in the kitchen,' he said ominously. 'And there's a new cook. What happened to the old one?'

I couldn't tell him the full story, he wouldn't have believed it. The old cook had put some pork chops in the oven with the plastic wrapping still on it. And she had cooked it that way. And the next day she got rather ill. She had puked up, her false teeth had flown out, and she had flushed them down the toilet.

'She...err...left.' I abbreviated my answer.

'Well, the new one is stark, raving mad!' said Mr Parker. 'I went in the kitchen and she started *blessing* me. Then she dropped a load of pots and pans, and started blessing them too!'

Again, I had to abbreviate my answer. I really should have suspected something when I first interviewed her. 'Why are you always smiling? I'd asked her and she had flashed me a crucifix.

'She's got religion,' I explained.

'She's got more than religion. She's stark, staring mad. First, she tried to convert me, then she turns to me and says: "Can you smell it?" And I say, "Smell what?" And she says, "The smell of sex! It's following me around!" And I said, "Well, I dunno what sex smells of, let alone sticking my nose in it!"

'I think she might be having some personal problems. She's not a bad cook.'

'I don't care what kind of cook she is. If she's going round smelling sex and turning my kitchen into a church, she's out!'

'Out?' Alarm bells were ringing in my head. 'As in sacked?'

'What the hell else do you think I mean? Get rid of her!'

'Erm...I can't,' I said weakly.

'Why not?'

'She's got a contract of employment. She might claim unfair dismissal.'

Mr Parker paused, a dangerous gleam in his eye. 'Oh yes, contracts of employment, that was your bright idea, weren't it? I dunno about you, Mister Kusy. I can't sack staff when I want to no more? You're giving me a right pain in the arse.'

I raised my voice in mild protest, but he wasn't having it. 'That's one strike against you, Mr Kusy,' he said as he swept out of the office. 'Two more and you'll be on my shit list. You do *not* want to be on my shit list!'

*

To calm myself down, I paid a visit to Betsy. Okay, I was risking being intimately frisked for her pocket money, but Mr Parker's visit had really unsettled me. I needed to be in the company of a friend.

'How are you today, Betsy?' I asked her.

A spidery little hand gestured for me to sit down, then settled on my knee.

'Never felt better in my life,' she said cheerfully. 'Mind you,' she continued in a confidential whisper. 'I am having trouble with my boyfriend.'

'Boyfriend?' I said, secretly relieved. 'I didn't know you had a boyfriend.'

'Yes, I'm sorry,' giggled Betsy. 'I should have let you know. You probably think me a two timing tart, but I'm not. That's him over there…Bertie.'

I looked up at the happily waving figure across the hall from us. He looked like Punch with a red nose.

'All the jolly!' Bertie shouted over, and lifted what appeared to be a tot of rum in our direction.

'Is he new?' I enquired. 'I haven't seen him before.'

'No, you wouldn't have done,' said Betsy. 'He's been away in hospital, having an operation "down there". But he says not to worry, he's still in perfect working order!'

'What's with the rum?' I asked. 'And why is he wearing a fez on his head? Does he want to be a magician like Tommy Cooper?

'Oh, he does like his rum,' giggled Betsy. 'And the fez is part of his most treasured memory – of the time when he jumped on a P & O ship bound for Cairo. "I'd just had a major operation for ulcers," he told me, "and the surgeon advised me to spend some time by the sea."'

'So what's your problem with him?' I asked. 'Why doesn't he come over here?'

Betsy tossed her tiny head. 'Oh, I had a tiff with him earlier. He's very set in his ways.'

'How's that, then?'

'I said to him, I said, "I feel wonderful! I think I'm in love!" and he said nothing. So I said, "Don't you want to know who I'm in love *with?*" And he said, "If you're thinking of sharing a room with me, you can't. You don't get up early enough." Well, that was rude, wasn't it?'

I nodded and laughed, and decided to change the subject.

'I'm so sorry you're having trouble with the Social. We're doing everything we can to sort that out. The Social can be very difficult.'

'Don't knock the Social,' said Betsy, suddenly serious. 'The Social is the best thing to ever happen to this country.'

'What do you mean?'

'I mean...Here, Ethel, wake up!' Betsy whipped out her hand fan and directed it into the left ear of her sleeping companion. 'Wake up and help me tell Mr Kusy about the Social!'

There was a pause as Ethel's ear began to turn blue, and

then a sudden gasp as her open mouth flapped shut and her eyes bugged out of her head. 'Jesus at Christmas!' she snapped grumpily. 'Did you *have* to do that?'

'Well, we don't want a repeat of last Thursday when Father Brown passed by and saw that gaping gob of yours and accidentally gave you communion, do we?' cackled Betsy.

Ethel harrumphed a bit, but then settled herself down and said, 'Are we back talking about the "good old days" again? Here, switch that machine of yours on again, young man, you'll have a book by the time she's finished!'

'We're talking about the Social,' said Betsy, ignoring the insult, 'and how hard life was for people before it. I mean, when we were growing up there was not much money about, was there? And most families were *big* families. Not like today. And the mothers were like really the *slaves*. They were on all day and all night. Sending the children to school, cooking and washing for them, sewing and cleaning, and all that lark. Didn't have washing machines like they got now. I was one of six kids. But I learnt my lesson. I only had one of my own!'

Ethel gave a reluctant nod of agreement. 'It was a saying in those days, that "to be poor, and look poor, was a damned bad habit." You see, the thing was, if you sunk or let yourself sink below a level, that was curtains. Because there was no social security, there was nothing. And no-one would help. You'd probably just die of starvation or cold in the winter. Because nobody was going to give you anything. It's like with a native dressing up in the jungle so that civilised people would give him half a coconut, cos he felt if he *didn't* he'd just end up lying down and dying of fever. I think that was how people were. They must, for their own self-respect, keep up their appearance. Because if you dropped behind, you'd had it!'

'There was nowhere to get help,' piped up Betsy. 'Nothing!'

'The most you got,' continued Ethel, 'was ten shillings a week, the old age pensioners. But if they had any sons or daughters – it didn't matter whether they were a long way away – they didn't get a penny, because their children were expected to support them. That was the system.'

I was finding this fascinating, and more than a little disturbing. Having been born the day rationing finished in the U.K. (17th June 1954), I had experienced none of this hardship. Nor had any of my generation.

'What happened, then,' I asked, 'to elderly people who had no-one to look after them?'

'They got 'em in the workhouse!' said Betsy.

'Yes, the workhouse,' agreed Ethel. 'That only stopped after the second world war. Many of the old hospitals today used to be workhouses. In which they put the men in one part, and the wives in the other, and kept them separate. And they may slag the Welfare State, but it has done away with that. This is what we all fought for.'

'What did they do in the workhouse?' was my last question.

'Well, nothing much,' mused Betsy. 'They took the old folk in for maybe a couple of years, and then they let them out – if one of them got a job, maybe.'

Ethel assumed a stern tone. 'This is what happened after the war, where everybody fought to ensure that never again would old folk have to suffer the indignity of the workhouses.'

'That's the way it was,' said Betsy. 'There was no *work* for anyone, especially in the winter. Builders, like my husband, didn't hardly work at all. The workhouses were always full in the winter. My father, who was a bricklayer, was always in and out of work. Until he got his job in Mitchells in Dulwich, where he was for 30 years. He only got that after the second war, when they kept building and building. That put everybody *in* work.'

'So that's it,' said Ethel in conclusion. 'People figured, after the war, that "if they can find the money to fight a war, they must have money to build us decent homes and our old people decent hospitals!" Which is how the Welfare State came in...'

I switched my Walkman off. This had been particularly interesting to me. Blow the petty officialdom which wouldn't top Betsy's fees up by a measly fourteen pounds a week.

I would never take the Social for granted again.

Chapter 8

Don't put your Director on the Stage

Christmas was upon us, and the home was alive with excitement and anticipation. Streamers and bunting were going up all over the place, and Miss Sherring, the residents' unofficial spokeswoman, was busy planning the big Christmas party in the visitors' room. I had a lot of time for Miss Sherring. Sharp as a tack and very self-motivated, she took a keen interest in all the home's activities and was to be found each and every day filling in her time with one thing or another – reading, doing crosswords, flower pressing, making things to sell at fetes and bazaars, or chatting to other residents.

'Do you play bridge?' she asked when I dropped in to see how she was doing. 'We need a fourth for tonight.'

'Well, yes I do, as a matter of fact,' I said. 'I played in the Evening Standard Bridge Congress a few years ago.'

Miss Sherring's blue rinsed hair shivered a little as she digested this unlikely piece of information.

'How did you do?' she said in a sceptical tone.

'Not very well. It was an all nighter and they wouldn't let me and my partner play two diamond Stayman because it was an American convention. We were forced to play basic Acol and came last. Oh, and in the final hand, I found myself playing against Omar Sharif and went four down in six no trumps.'

'So, you're not very good, then?' Miss Sherring's face visibly brightened. 'Well, you *must* play with us!'

I returned to my office to find a growing line of residents waiting outside. They all wanted the same thing: Paracetemol tablets. 'It's that Bill again,' complained Mr Reitz. 'He's been shouting all morning – we're all sick in the head!'

Yes, I had heard a bit of that shouting while I was with Miss Sherring. What's Bill upset about now?

'Oh, he came across the Supervisory Report Book in Matron's office,' said John Gray as he lounged in to help me distribute painkillers. 'And he read all the recent reports of his bad behaviour. Confronted with his unpopularity in black and white, he kicked up an awful stink and ran amok in the dining room at tea-time, chasing care assistants round the tables with his white stick. His reign of terror only ended when he tried to hide in the toilet when deaf old Mrs Glendening was already in it. In the ensuing scuffle, both parties fell down and couldn't get up again!'

*

As for me, I was engaged in a two-pronged attack in trying to regain the favour of Mr Parker.

First off, I got rid of the cook. 'How are we today?' I asked her over lunch, and when she came back with: 'I think I can smell shit – this smell of shit is following me around everywhere I go!' – I sniffed the air and said, 'Yes, I can smell it too. Awful, isn't it?' Overjoyed at being understood for once, she then confessed: "The bishop from my old church is following me around all the time. Should I be scared?" Which was my cue to look over her shoulder and say: "Oh yes, there he is. And he's holding an axe.'"

She fled screaming.

My second gambit to make a good impression wasn't half as successful. Every Christmas, before their big party, the residents were treated to a 'show' put on by the staff and their friends. This year it was the 1920's Broadway classic 'No, No, Nanette', and in my new capacity as the home's Director I was encouraged – no, *expected* – to be in it. I wondered why Matron hadn't volunteered to be in it too, but I didn't have to wonder long. 'Oh no, dear,' she said when I posed the question, 'I don't do theatricals.' I rolled my eyes at that. Matron was all about theatricals.

My own experience of theatricals was limited to a single role in a nativity play at primary school. I was playing one of the Three Wise Men – Melchior, I think – and my mother had made me a beard out of horsehair pulled from the seat of an old chair. It irritated my face so much that when I wandered onto the boards, scratching it fit to burst, I wandered straight off again...and fell into the orchestra pit. Baby Jesus was not amused at being upstaged like this, and neither was the school principal. I had never been invited to play anything again...until now.

The wardrobe department did me up a treat. I wasn't sure about all the rouge they slapped on my cheeks, but the stripy seaside jacket they found for me and the flowing beige trousers (with braces) were extremely to my liking. As was the fetching boater hat, which I adjusted to a jaunty angle for best effect. To top it all off, I had shaved my beard and was now sporting a thin 'period' moustache, which I thought made me look rather dashing.

Someone in the audience did not agree. The moment I entered stage left and prepared to issue my opening line, a loud voice at the back of the hall shouted: 'Oi, Mr Kusy! You look like a ponce!'

There was short, shocked silence and then everybody's heads turned round to see who the perpetrator of this statement might be. And there, hugging his sides with laughter, was Old Bill. 'Well, he does, don't he?' he howled unapologetically. 'He couldn't look more like a ponce if he tried!'

Then, as everyone's eyes swivelled back to me again, I did what any amateur thespian confronted by a master octogenarian heckler would do.

I froze.

They say that everyone has a secret fear – one thing that might happen to them which would dwarf all others in its scale and sheer, unadulterated awfulness. For me, it was public embarrassment. It had taken all my courage to get up on that stage and perform, and now I was stuck up there – all on my own – and I could not remember my lines. I looked down and saw my mother, her mouth open in mute apprehension, and next to her, Mr Parker glowering with undisguised annoyance. Behind them, an anonymous sea of faces opened up to me, all of them shifting ever more uncomfortably in their chairs.

There was no more time to waste. I had to say *something*.

'I must call Lucille on the *telephone!*' I announced importantly. But there was no telephone in sight. Worse still, there was also no sight of the chorus line of girls who should have been dancing in with the opening number of "Flappers are we".

'You're in the wrong act!' hissed the distraught director of the show from the wings. *'You're in Act three!'*

Oh dear, it wasn't my night. If I went on like this, it would be a very short play.

'Yes, I must *find* the telephone!' I announced even more importantly…and scuttled off the stage.

The cast were very forgiving, I must say. The play was retrieved, and we all got a very big applause at the end. But I had my eye on Old Bill. If I had my way, he would not be coming on any more public outings for a very long time.

Back at the home, and having promoted the nervous young assistant cook, Miss Mumuni, to the dizzying heights of first cook, I sought out Betsy and Ethel again. They were seated this time in the canteen and they were sharing a pot of tea and a Battenburg cake.

'Sorry to interrupt you, ladies,' I said as I hovered around them. 'But I was wondering, how was Christmas in your day?'

Betsy peered short-sightedly up at me, then down at my trousers. 'Where is it, then?' she mumbled through her cake. 'Get it out!'

I looked down in confusion. I hoped she didn't mean what I thought she meant.

'That bulge in your left pocket,' she giggled. 'Or are you just glad to see me?'

'Oh, you mean my Walkman,' I said, inwardly relieved. 'Do you want me to tape you again?'

Betsy's eyes lit up as she spotted the blinking red light of the recording device. 'Ooh, you can tape me wherever you like. I'm open to suggestions.'

Pretending I didn't hear that, I repeated my question. 'How was Christmas when you were young? Did you have a rich table of food, and lots of things that you didn't have the rest of the year?'

'Hmmph!' snorted Betsy. 'It didn't mean much to me!'

'It didn't mean much to any of us!' said Ethel. 'No! We might get an orange, and perhaps an apple…sometimes a few coppers…we didn't get *toys* or anything. Our parents couldn't afford it!'

'Yes,' agreed Betsy. 'We had a stocking with an orange in it…perhaps a bar of chocolate…and a picture book. That was your Christmas.'

'What about the Christmas turkey?' I asked.

'No…no,' quavered Betsy. 'You were lucky if you even *saw* a turkey.'

Ethel shook her head. 'I was fortunate. I lived in Bloomsbury and everybody there went to Leadenhall or Smithfield markets. And late on Christmas evening, they auctioned off all the leftover turkeys cheap. So it *was* possible to be poor and have a good Christmas table.'

'You? Poor?' scoffed Betsy. 'Don't make me laugh! Living in Bloomsbury and calling yourself poor?'

'No, we had it rough too,' protested Ethel. 'Especially in

the twenties, when nobody had any work. I can remember – when I was twelve, it must have been 1922 – women coming round with a bucket and offering to scrub your doorstep. The front steps, that is, which was always washed in those days.'

'My mother and I used to do that,' said Betsy, giving Ethel a glare to suggest it was all her fault. 'We went round knocking on doors, and offering to "half stone" the steps for sixpence.'

Ethel looked defensive. '...And I can remember many a time my mother giving this woman a shilling – and we *weren't* well off – because, as she said, if she didn't give this woman a shilling, probably her kids weren't going to eat that day. And I'd come home and perhaps there'd be one sausage on my plate instead of two, because this woman had got her shilling!'

'My mother,' said Betsy, 'she took in all the gentry's washing from Beckenham. Yerse! She'd be up half the night to get the washing done, and dried, and taken back – to get the money to get the six of us kids a meal. That went on till 1945, when the building trade picked up and my dad he went to Mitchells.'

This sounded familiar. When my dad died young, my mother had also been forced to darn dresses at night in addition to a day job. Many a night I had drifted off to sleep to the sonorous hum of her Singer sewing machine.

'So,' I said, changing the subject, 'are you looking forward to the Christmas party?

'I should coco,' cackled Betsy. 'And I'm bagging Bertie for all me dances. Nobody can trip the light fandango like Bertie.'

'If he doesn't fall over,' observed Ethel sourly.

'Oh, don't you worry about Bertie,' said Betsy with a titter in her voice. 'He's the only one who doesn't get pissed at Christmas. All the rest of us get tiddly and giggle away like maniacs the whole night. Bertie can drink *anybody* under the table.'

'When did he start drinking rum?' I enquired curiously. This Bertie sounded a right rascal.

Betsy and Ethel went into a quick whispered huddle. The jury was evidently out on Bertie with them too.

'Well, he *tells* us it was when he was in the Navy,' said Betsy at last. 'But we can't work if he was in the Navy before the Air Force. Or since. Or at all. But we're beginning to think he was never in any of these things.'

I laughed. 'What wars does he say he's been in?'

'Oh, most of them,' puffed Ethel. 'Not only the first and second world wars, but the Spanish Civil War, fighting alongside Mussolini in Uruguay, and even in Vietnam. He also throws in the Boer War occasionally. If he listened to the news, he'd probably be in Afghanistan fighting the Russians!'

Whether by chance or by accident, the very next person I came across, as I proceeded down the corridor selling Christmas Dance tickets was the man himself, Bertram Button.

Bertie was holding court at the mobile library, flourishing a book about Rommel in Africa and telling his awestruck assembly of ladies: 'I was there! I was Montgomery's batman! Oh yes, Monty and me go way back. It was just before we knocked back those Zulus!'

'Oh, how exciting!' fluttered elderly Mrs Winch. 'Was that when you were a bomber pilot or a submarine commander?'

'Or in the wrong century,' I politely interrupted. 'Didn't we fight the Zulus sixty years earlier, at Rorke's Drift?'

'What do you know, young man?' said Bertie, looking a little caught out. 'There's Zulus all over the place. I even saw one at Clapham South tube station the other day!'

I opened my mouth to comment but then closed it again. Bertie had come over all strange. His hands were shaking, a thin dribble of spittle had appeared at the side of his mouth, and his legs looked like they were about to give way.

'Here, let me take him,' said John Gray, fortuitously materialising at my side. 'What's the matter, Bertie? Haven't you been taking your pills again?'

Bertie leant on John's shoulder and fumbled around in his pockets. 'I knew I was supposed to remember something,' he murmured, pointing to a knot in his handkerchief. 'I just forgot what it was.'

'Well,' said John in a kindly voice. 'It won't get any better. Not with Parkinson's. But you're off the eye drops, so that's an improvement, isn't it? I think you're doing very well. And keep having that can of brown ale a day. That's going to do your bowels no end of good.'

'Stuff your can of brown ale,' muttered Bertie as he sank uncomfortably into a chair. 'What I need is a tot of rum. Yes, the world would be a much better place if everyone had a tot of rum…'

Bertie was quite a different man now, all the boastful stuffing knocked out of him. I gestured for John to leave and put a comforting hand on his shoulder.

'What's your story, Bertie?' I asked. 'How did you end up here, and how did you feel when you were first admitted?'

'Well, I'd just lost my wife,' he said, holding his own hands together to stop them trembling. 'And how did I feel? Pretty disjointed, really. When you've been together with someone for 50 years, and then suddenly get separated – not knowing what to do, where to go, who to turn to – your life pretty much falls apart. I'm not sleeping well, you know…disjointed. They tell me off for wandering a lot at night, especially into other residents' rooms and giving them "frights", but I can't help it. My life is disjointed. I don't know whether I'm coming or going, most of the time.'

I struggled to think of a way of cheering Bertie up. A big man with a wide barrelled chest and no neck to speak of, he seemed small as a child now. My heart went out to him, and I found myself chanting inside for his happiness.

'But you're a hit with the ladies,' I ventured. 'They all dote on you!'

Bertie gave a low sigh. 'Well, I've got a good front, and it's all women here, isn't it? Don't see many men. And the staff, well, they mean well, but they're too busy to sit down and chat. I understand that. Of course, I've got friends from the outside…relatives…my old neighbours. But they promise to come and see me, and I sit and wait for them to come and see me, but they *don't* come and see me. My trouble is, I never made my fortune. If I'd made my fortune, well…'

Bertie reminded me of my grandfather. A once proud man laid low by the ravages of time, trying to keep his dignity in the face of diminishing health and extinct opportunities. Over the next half hour or so, I learnt that he had been a baker by trade, and that his sudden, impulsive leap onto that P & O ship 50 years ago had been the one brave thing he had done in his life. All the rest of his heroics in various world conflicts – in typical Walter Mitty fashion – had been the invention of a fevered, and very regretful, imagination.

But this particular Mitty had one more string to his bow.

And when he drew it, all hell would break loose…

Chapter 9

Bertie Steps Out

I returned to the home after the Christmas break to find it in an uproar. Three police cars were parked outside, and inside, half the committee and all of the staff were running around like headless chickens.

'What the matter?' I asked John Gray. 'What's going on?'

'Bertie's gone missing,' said John. 'Actually, he went missing three days ago, but Matron was holding the fort while we were all away and she didn't notice his absence. Then the silly woman rang the police instead of Mr Parker and now we got a media storm on our hands: the papers have us down as "the care home which lets its residents out to almost certain death in the snow".'

This sounded grim. 'What's Mr Parker's take on it?'

'Oh, he's absolutely furious, wants Matron's guts for garters. But he can't have them.'

'Why not?'

John's bushy moustache twitched in an amused smirk. 'Because Minnie Glendening put her in hospital.'

Minnie Glendening? Surely not. A kinder, sweeter old lady I had yet to meet.

'I just heard from John Trundle,' said John. 'It was the middle of last night, and Minnie was in one of her moods. So she got it into her head to get some attention, and started banging on the radiator with her stick until the night care assistant arrived. Then she started off with her: "I can't see! I can't hear!" But she could see alright. And a few minutes after the assistant had got fed up and told her firmly to go back to sleep, she was up and off and found her way upstairs, shrieking "I want to see Matron!" until she arrived outside Mrs Butterworth's door. Then she banged on the door until Matron woke up and came to see what all the fuss was about. She opened the door to an irate Glendening, and came out with a haughty "Don't be silly, Minnie! Don't be silly!" Which of course really got Minnie's back up. Suddenly, she lunged out and grabbed at Matron with both hands, grunting: "Who is it? Who is it?" Mrs B. was petrified, and tried to fend her off with "Let me go, dear! Let go!" But Minnie had her in a grip of steel and pinned her to the wall by her throat. Finally, the assistant came and prised Minnie off, but she lashed out with one hand and cracked Matron across the chin with one mighty wallop, and put her in hospital with concussion.'

'Blimey,' I said, 'Every day a drama. So what's happening now?'

'What's happening now,' said John, 'is that there's a nationwide police hunt for Bertie. We've tracked his movements to Wandsworth, where he apparently borrowed two hundred quid from a friend. Oh, and he picked up his passport. I can't imagine what he would want with a passport, can you?'

A shiver of apprehension ran through me. And the germ of an idea of where Bertie might be planted itself in my mind.

'Oh dear,' I thought to myself. 'I hope I haven't reignited his travel ambitions.'

Memories of our meeting at the mobile library flooded back to me. Bertie had given me a vivid account of his times in Cairo, and I, to be friendly, had given him a vivid account of my times on an Israeli kibbutz. To be fair, my times on the kibbutz sounded a lot more fun than his in an Egyptian hospital and when Bertie heard that everything had been free there – food, booze, accommodation – his OCD kicked in and he'd begun plaguing me with questions of how he could go to one. I hadn't taken him seriously, of course, so had strung him along with a series of non sequiturs like 'Maybe tomorrow, Bertie,' or 'Okay, Bertie, we'll see what we can do.'

The last day before Christmas, however, it all got too much. Mr Parker had been on my case to finalise his 60-table Supper Dance down at Wandsworth Civic Suite, Miss Mumuni, the new cook, had collapsed under the strain of having to cater for 200 people going to the residents' party, and Matron was fussing about being...well, Matron. All I needed was for Bertie to turn up with one more whining nag about how he could visit a Middle Eastern working commune.

'Oh, for God's sake, Bertie,' I'd snapped in a moment of ill-advised temper. 'If you want to go to a kibbutz so much, *go!*'

My mind fled back to the last time I had issued such unwise advice. A few years earlier, I had gone to Norwich to train as a Careers Guidance Officer and my first videotaped interview, with a spotty little teenager called Duane, had not gone well. 'So,' I'd asked Duane, 'what do you want to do then, when you leave school?' I wasn't expecting much: Duane's school achievements amounted to a single CSE qualification in metal work. But Duane had surprised me. 'I want to be an astronaut,' he said chirpily. And I had replied, on camera, 'Yeah, well, why not? Go for it!' My tutors had failed me without blinking.

Now I was facing a similar disaster. Though maybe I was imagining it? Yes, maybe Bertie had taken refuge in some safe house and was wandering his way back to us right now.

No such luck.

Two days later a blue aerogramme with a foreign stamp fluttered into my post box. The sender's name was marked as Mr B. Button, and below it, penned in the same spidery script, was a familiar address: Kibbutz Ein Harod/Meuchad, Israel.

Closing the door to my office, and then locking it, I carefully opened the letter and read its awful contents:

Dear Mister Kusy, I took your advice. The kibbutz is great. They got me cutting vegetables and everything. Only downside is, they haven't got no rum, only Sabbath wine. Staying here for a bit, then on to see that Moses mountain you told me about. Ta ra, all the jolly! Bertie.

My first reaction, unexpectedly, was: 'Good for you! About bloody time you did something instead of just talking about it!' But then reality set in. What was a 75 year old man with Parkinson's disease and hardly any money going to do in a strange country where few people spoke English and where no-one knew of his condition?

Then I noticed the small passport sized photograph that had fluttered out of the letter. It was plastered to the top of my shoe. Bending down to retrieve it, I recognised Bertie, flanked by two grinning children, standing in what appeared to be some kind of Arab bazaar. He was wearing his fez and his woolly dressing gown and his carpet slippers. He could have been one of the locals. But it was the expression on his ruddy, red-nosed face that struck me. Now he didn't just look like Punch. He looked as pleased as Punch.

In a panic, I rang Brenda.

'One of my old gentlemen has gone crazy,' I told her, 'and

hopped on a plane to an Israeli kibbutz. I don't how they let him on, dressed in all his night clobber, but they did. What do you think I should do?'

'Good Lord, Frank,' said my old editor friend. 'What is it with you and difficulties? Did that Buddhist meeting inspire you to seek them out?'

I sighed. 'No, but they keep finding me anyway. And this one is a doozy. If this guy pegs out on me 4000 miles from home, I'll never forgive myself.'

'Well, you're doing the right thing, Frank,' said Brenda in a comforting tone. 'You're taking responsibility. Buddhism is all about taking responsibility. Have you thought about phoning the kibbutz?'

'They haven't got a phone,' I responded. 'Well, they didn't when I was last there. All I've got is the fax number of their representative in Jerusalem.'

'Well, drop him a line,' suggested Brenda. 'If that doesn't work, how about writing a letter back?'

'You don't know this guy,' I said. 'He's a persistent wanderer. By the time any letter reached him in Israel, he'd probably be in Egypt.'

'Well, there's only one answer, isn't there?' said Brenda at last. 'If you don't want the police involved, and a potential international incident, you're going to have to go over there and bring him back yourself.'

I began ringing round travel agents.

*

Twenty-four hours later, and having put John Gray in charge of the home while I was away on a "family emergency", I touched down at Ben Gurion airport. The dull, acrid reek of tarmac mixed with the sweet, more flavoursome, aromas of the Promised Land to set off a trail of memories as I stepped off the plane. Had it really been five years since I was last here? It seemed like only yesterday. The warm January breeze fanned my face as I recalled the 'Project 67' organisation which had brought me here and the last-minute letter I'd had from them informing me that our group would not be going to Ein Harod at all, but to some other kibbutz to the far north of the country. How lucky I had been to stubbornly ignore this letter and proceed on to Ein Harod anyway. It let me and two others in, and I enjoyed three of the happiest months of my life on one of the best kibbutzim in the country, complete with a swimming pool, a brand-new cinema, and a concrete bunker disco. All the rest of my group had gone to the Golan Heights and lived in fear of being bombed by Arab terrorists.

Past airport security, I was welcomed into the country by fat, smiley Nelson Ben-Ami, the Project's Jerusalem representative.

'Ah, you come back to us!' he said, giving me a big bear hug. 'Did you bring Johnnie Walker Red Label?'

I handed over the promised bottle of the universally popular whisky, and – forgoing the usual formalities – showed him the picture of Bertie.

'I'm sorry, my friend,' I said urgently. 'But I must find this man. Have you seen him?'

'Have I seen him?' bellowed Nelson, his lips parting to reveal a cavern of nicotine-stained teeth. 'This man is famous. He is best friend of our Moshe Dayan! He help him defeat Egypt and Syria in Six Day War!'

My eyes rolled in my head. 'Is that what he told you?'

'It is true! It is so! And now he is special undercover operative for our Prime Minister, Menachem Begin!'

This really was too much. Even for Bertie.

'Undercover operative?' I said. 'How can you be so sure?'

'I find Mr Bertram in Arab quarter of Jerusalem,' said Nelson patiently. 'Everyone is talking about him because he is drinking whole bottle of our Arak brandy without blinking. And his clothing is most cunning. What other Englishman would dress in such a fashion? Yes, he is like your James Bond – shaken but not stirred!'

'So where is he now?' I asked, trying to keep the scorn out of my voice. 'Is he still in Ein Harod?'

'Oh yes,' beamed Nelson, pointing me to his car. 'I will take you to him immediately!'

It was gone noon when we arrived at my old stamping ground in the Yizreel Valley. The rains had just come and the place looked pretty deserted. Still, I could see the olive groves and grapefruit orchards where I had been forced – no, *encouraged* – to climb up 20 foot ladders every morning. Ah yes, many a happy time I'd had up those ladders – fending off three-inch tree frogs, trying to avoid half foot butterfly cocoons – with manic work boss Svigi chivvying me on to ever more dangerous branches.

Svigi was in fact the first person I met inside the kibbutz compound. Wearing the same dusty, grey work outfit and the same sour expression as before, he did not look thrilled to see me. 'I'm back!' I said cheerily. And he said, 'Ah yes, the volunteer who does not like to work much.' He had evidently not forgiven me for screwing up his precious Israeli work ethic which determined that all volunteers had to work 6 hours a day, from 6am till noon, with no incentive at all to fill their quotas. It had been the height of the grapefruit season, and his usual 'encouragements' for the volunteers to pick extra hard had fallen on deaf ears. 'Why don't you try a bit of capitalism,' I suggested, 'and let us go home early once we've filled our baskets?' And he had, and everyone had picked with gusto and filled their baskets in only 3 or 4 hours. But then the kibbutz managers had found out what was going on and (I felt really bad about this) had relegated Svigi to the Plodot steel factory.

Svigi was not the only person not enamoured to see me. Over in the corner, as I sat waiting in the bright, spacious dining hall for Nelson to go get Bertie, I spotted Nathan, my original work boss. Small and quietly furious, he was glaring at me as though I just had murdered his children. Which was a trifle unfair, since it was he who had made my very first day as a kibbutnik so unpleasant. He had driven me and a Swedish girl called Chris to some heat-blasted mountain on the Lebanese border, packed us into thick cotton suits (with helmets) and told us we were going to be dealing with...bees. Yes, thousands of angry bees. Which got *really* angry when we smoked them out of their hives to get at their honey. Nathan couldn't understand it when I picked up the first hive to put in his truck and then suddenly put it down again. 'Work!' he screamed hysterically. 'Why you no work?' The reason was simple: three bees had found a hole in my trousers and were now buzzing around my loins. 'Problem!' I shouted back. 'Big problem!' But Nathan didn't care about my problem, just started jumping up and down and cursing me with ill-disguised venom. It was then that the first bee stung, and I lurched forward and vomited into my helmet. 'Oh dear,' I thought, 'I really shouldn't have had semolina for breakfast.' Then the second sting whapped into me – rather too close to my left testicle – and I went into anaphylactic shock. Poor Nathan. He'd had to forget all about his precious hives and drive me two hours to the nearest hospital.

Yes, I was not the most popular bunny on the kibbutz. But I was not here to be popular. I was here to find the biggest King Liar in the Middle East and restore him to his rightful throne in Clapham.

This proved to be much more difficult that I thought, however.

'Bertie gone,' pronounced Nelson, returning from his tour of the living quarters.

'Gone where?' said I, feeling my prey slipping from my fingers.

'He go Mount Sinai,' shrugged Nelson. 'Some volunteers just put him on bus.'

Sinai? This was exactly what I didn't want to happen. I knew, from Bertie's letter, that he wanted to see the mountain where Charlton Heston had given the Ten Commandments to millions of movie goers, but why now? Why couldn't he have waited just one more day and saved me a long, arduous exodus into present-day Egypt?

'I have a very big favour to ask,' I told Nelson, looking at my watch. 'I don't know when the next bus to Sinai will be, but it will probably be hours. Can you run me down there? I have to find Mr Bertie tonight!'

And it was true. I had already spent half of Saturday on foreign soil. If I didn't get Bertie back home on the Sunday night flight, and report for work the next morning, I'd have a very suspicious Mr Parker on my hands. I was also desperately worried for Bertie's health. He had been off his medication for almost a week now: I shuddered to think what would happen if he tried to climb a 7000 foot mountain.

Nelson mulled over my question a long time, rubbing his hands over fat, jowly cheeks as he lost himself in thought.

'I like to help Mr Bertie, of course,' he said at last. 'But maybe he is on "secret mission", maybe he does not *want* to be found!'

I sighed. This was getting ridiculous. A very special incentive was required.

'Will *this* help you to want to find him?' I said, reaching into my rucksack and withdrawing another bottle of Johnnie Walker. 'This is my own personal stash – Blue Label. Much better than Red.'

Nelson's eyes bugged out of his head. '*Blue* Label?' he gurgled in disbelief. 'I have heard of this thing, but can only dream! You give to me? Thank you! Wait here, I start the car!'

*

We drove in a straight line south – passing Jericho, Bethlehem and Beer-Sheba – until we came, around dusk, to the emerging coastal resort of Sharm el-Sheikh. Then, after a short pause for gas, food and supplies, we pressed on – in pretty much pitch darkness – for another three hours until we came to a series of mountains, one of which enclosed an ancient monastery.

Fortunately, the moon was coming up, so we began to have some visibility. Even more fortunately, not ten minutes from leaving the car, I found a series of steps leading up a mountain.

'Wow, this is a stroke of luck!' I told Nelson. 'This must be those "Steps of Repentance" I read about. There's about 3000 of them and they go all the way up to the top of Sinai. Follow me!'

I could not suppress a feeling of excitement when I started climbing that stone staircase. It was a dream I had had five years earlier – a dream which had been smashed when I ran out of money at Sharm el-Sheikh and had to turn back to the kibbutz. Indeed, I had been so strapped for funds that I'd had to sell three pints of blood in Jerusalem to afford the final bus fare.

About half an hour later, my excitement turned to puzzlement. The steps just stopped.

'There's no more steps!' I wailed down to Nelson. 'What's going on?'

Just then, as the moon came out from behind a thin veil of cloud and bathed us in the full force of its ghostly glow, an even more ghostly voice shouted across to us from the adjoining mountain.

'Helloooo, Mister Englishman!' it said. 'You are on ze

wrong mountain!'

I looked over and there, not half a mile away, were a group of German hikers. And bobbing up and down on a camel between them was a familiar-looking fez.

'It is Mr Bertie!' said Nelson, clapping his fat hands together in joy. 'And he is riding, how you say, "in style!"'

'But there are steps on *this* mountain!' I shouted back at the Germans.

'Ho, ho, ho!' boomed their annoying spokesman. 'But not enough steps, ja? Now you must grow wings and fly up ze mountain!'

How was I supposed to know that the 6th century monk who'd chipped his steps to the top of Sinai had a disciple? Or that this disciple had decided to chip his way up to the top of a different mountain? Or that said disciple had died before he had completed not a third of his task? All these things I learned when, having clambered down the *faux* Mount Sinai, I engaged two camels – with knowledgeable Bedouin guides – to take us up the real one.

I was not in a good mood.

*

'Oh hello, Mr Kusy,' said Bertie sleepily. 'Fancy seeing you here.'

My red-nosed quarry was sitting on a wide, flat rock with what promised to be a perfect view of the coming sunrise. He was swathed in warm blankets against the freezing cold and was nursing a small pot of tea, along with a plate of biscuits.

'How did you manage those last 700 steps?' I puffed at him. 'Did you tell someone you're a direct descendant of the prophet Isaiah and get a piggy back?'

'That's what I like about you, Mr Kusy,' said my fez-headed fugitive. 'You always see the funny side of things. Do you want a biscuit?'

I shook my head and flopped down beside him. 'I'm not feeling very funny, Bertie. Do you know how much trouble you've caused? I've been worried sick!'

'And you came all this way to see me. And Nelson too. How nice.'

'It is nice to see you too, Mister Bertie,' said Nelson, bending low and looking conspiratorial. 'I hope we have not interrupted any "secret agent" activities?'

'Blow the secret agent activities,' I muttered, rooting around in my bag and fishing out a pot of pills. 'Here, take a couple of these, Bertie. You're not looking well.'

Bertie's hands were trembling as he took the proffered medicine. 'Yes, I did have a bit of a turn earlier. Thank Gawd those German lads turned up and plonked me on that camel. I was in a right state. But I got to tell you this – I don't feel disjointed no more. I'm sorry I caused you trouble, Mr Kusy, I know I did wrong, but I wouldn't have missed this for worlds. Look…the sun is coming up!'

And indeed it was. As the first rays of a new day dawned, we were treated to a majestic sight – the mountain took on an almost ethereal glow and one by one all the lower crags and hills below were illuminated and then thrown into sharp relief against the desert floor. It was, an unforgettable and moving experience, and looking over to Bertie, I could see that he had tears in his eyes.

'So this is where God spoke to Moses,' he murmured. 'I can truly believe it.'

Nelson and I nodded, and for a few fleeting moments all three of us were joined in a silent, holy communion.

*

One slow camel descent and a frantic, seven-hour car dash later, we were back at Ben Gurion airport and praying that we could get two tickets on the night flight to London. Luckily, it being low tourist season, we did. Even luckier, nobody questioned Bertie's appearance. With his dressing gown, pyjamas and slippers in my bag – cleaned and neatly pressed back at the kibbutz – he was now wearing a Bedouin scarf, a much too tight pair of my spare jeans and a T-shirt saying 'I love Ibiza'. That, coupled with the fez and the mad, beaming expression on his face, made him stand out a mile!

'Goodbye, my friend,' said a tearful Nelson, clasping Bertie to his bosom. 'It has been an honour knowing you. And please, here is my address. So you can send me that signed photograph of you and our Golda Meir, remember?'

'That one happens to be true,' Bertie said with a tired wink. 'She came to my bakery in 1969, on her way to meet Harold Wilson in London.'

It was 2am on Monday when we touched down in Heathrow, just enough time to sneak Bertie back into the home and into his night clothes before the day staff arrived.

'Phew, job done,' I thought as I snuck back to my own home for a few hours sleep. 'I've returned the lost sheep to its flock without Mr Parker, its bushy-browed shepherd, being any the wiser.'

How wrong I was.

Chapter 10

Phoney Pheasants and Plasticine Pigs

'Ah, there you are, Mr Kusy,' purred a familiar gruff voice when I entered my office later that morning. 'I've been waiting for you.'

There, sitting at my desk, was Mr Parker and he was waving Bertie's blue aerogramme from Israel in the air. He was also wearing an expression that I had never seen before...sort of *conflicted*.

'I don't know whether to kick your arse or give you a medal,' he continued. 'Which do you think I should do?'

My mind raced. How stupid I had been to leave that aerogramme behind. I should have remembered that Mr Parker had a spare key to my office.

'Erm...let me explain,' I began. 'None of it is Mr Button's fault. He didn't...'

'I don't want to know,' said the Chairman tiredly. 'I've just spoken to Mr Button and I don't want to know any more of it. All that I *do* want to know is how we proceed from here – do we put you and that drunken old fool in the papers and make you local heroes, or do we put a quiet statement out that he wandered back here after a few days down a pub in Hackney?'

'I like the pub idea,' I said. 'It's far more believable than the truth.'

Mr Parker lifted his glasses, wiped his little piggy eyes, and blew his cheeks out in resigned surrender. 'Okay, it's your call, Mr Kusy,' he said. 'But I got to tell you this – if you open your gob and have one more of my residents gallivanting off like Marco Polo, you won't just be on my shit list. You'll be on top of it.'

Unsure whether to feel relieved or concerned at this outcome, I popped upstairs to the residents' rooms to see how Bertie might be doing.

He was doing fine.

'Ooh, you caught us at it!' giggled little Betsy. She was sitting on the bed next to Bertie, her hands intertwined with his and her head on his shoulder. 'Bertie's just telling me about his adventures. Did he really meet Lord Nelson in Jerusalem?'

I chuckled. 'No, it was *Admiral* Lord Nelson, wasn't it, Bertie? And we sailed down the Nile and took on more of those Zulus. Don't you remember?'

'Oh, yes, that's right,' muttered Bertie, pulling out his handkerchief and staring at the knot in it. 'I just forgot the finer details.'

It was so cute, these two lovebirds shifting about on that bed like awkward teenagers. I decided to make a discreet exit.

'Good to see you doing so well, Bertie,' I said, reaching for the door handle. 'Let me know if you need anything. Some food perhaps?'

'No, we're alright,' said Betsy happily. 'Mrs Caitlin's bringing us a pheasant.'

I stopped in my tracks. 'A pheasant?'

'Yes, she asked us: "Do you like pheasants?" And I said, "Yes, I *love* pheasant." So she leant up close and said: "Don't put it about, but I've got a dozen pheasants. In the deep freeze. They're really meant for the Jews Home, but you can have them. I can carve 'em." Wasn't that nice of her?'

This sounded a bit too nice, so I decided to investigate. And found elderly Mrs Caitlin making the same hushed offer of pheasants to Miss Sherring in the dining room.

'What's this about pheasants?' I said loudly, and Mrs Caitlin looked up startled and realised she was in trouble.

'Okay, you can have a pheasant too,' she said rather too quickly, then tottered off in a huff.

A few minutes later, she was back again. 'Look, about those pheasants,' she muttered conspiratorially. 'My daughter just called. She used the pheasants for a party. You're out of luck with the pheasants. But I've just thought...I've got a whole pig.'

'A...whole...pig,' I echoed flatly.

'Yes,' said Mrs Caitlin with an air of supreme confidence. 'You can have that. But on the strict understanding that you don't let it get out I'm giving it you.'

There was a pause as I stared her down, and then she relented.

'On second thoughts,' she said, 'let's play it safe. I'll just give you a *leg.*'

*

Outside of the home, I had to admit it, my social life was close to zero.

While important world events like Michael Jackson's *Thriller* album being released and Bob Marley getting his own postage stamp in Jamaica played out around me, I was popping one more 50 pence coin into the gas meter of my cold, dingy flat and tuning in to *Dallas, Cheers,* and *The Sale of the Century* on the three measly channels of my rented television.

Once a week, I would drag myself down the local Leisure Centre and play badminton. Once a month or so I would make a similarly dispirited visit to the Battersea Arts Centre and hop up and down to a punk or New Romantic band. But the friends I shared these rare activities with were not real friends, I felt little in common with them.

So I began to attend more Buddhist meetings.

The first local meetings I went to took place in the house of a tall, craggy-faced individual called Nick. I wasn't very impressed by Nick to start with; he talked too much and had a slightly patronising attitude that I found irritating. But then he told me something that did impress me. He had a helicopter.

'A helicopter?' I said, amazed. 'Did you chant for that?'

'Oh yes,' he said. 'Though I don't chant for "stuff" any more. I chant for enlightenment.'

'Enlightenment?' I said, feeling slightly provocative. 'How's that better than getting "stuff"?'

Nick eyed me warily. Then he told me his story.

'When I first started chanting,' he said, 'I imagined that enlightenment was this cosmic thing involving floating around on clouds, appearing anywhere at will in the universe etcetera. Sadly, I soon realised it wasn't about this at all. Instead, it appeared to about getting the things one wanted, so I spent a while chanting to fulfil my desires. Later on, it changed again, and I was told to chant about getting rid of my difficulties, which was yet another interpretation of enlightenment. Finally, in the last chunk of my practice, through reading the guidance of our mentor in Japan, Daisaku Ikeda, enlightenment has come to seem much more to do with achieving an unshakeable, indomitable life-state – where nothing can rock or disturb me.'

There was a brief pause as Nick nodded sagely to himself and I absorbed his words of wisdom.

Then I said: 'So, how did you get the helicopter?'

Poor Nick. He really did have his work cut out with me. The Jesuits had trained me well. I knew all the awkward, rhetorical questions like 'Why is God?' 'How tall was Jesus?' and 'How many angels *do* dance on the head of a pin?' All that was required was for me to pretend to be stupid and see how far I could go before Nick lost his precious indomitable life state and went for my throat.

But he surprised me.

'I was a lot like you when I started into this Buddhism,' he said with a chuckle. 'But then I met Richard Causton, the leader of our small Nichiren movement in the U.K. Dick is great. I never respected any guy – especially one in a position of authority – before. He changed my mind. You should meet him.'

'There's a lot of "shoulds" coming my way,' I said, sidestepping Nick with practised ease. 'I thought there were no rules in this Buddhism, but all I'm hearing is: I *should* chant more, I *should* go on courses and to more meetings, I *should* get a Gohonzon...'

'Oh, you should definitely get a Gohonzon,' interrupted Nick. 'And there's only one rule in Buddhism – to respect Life.'

'Is that why that girl in the meeting chanted for her dad to die?' I enquired rather sarcastically. 'Who told her to do that?'

Nick's face twitched. I could see his halo beginning to slip.

'Dick gave her that guidance,' he said with ill-disguised annoyance. 'And it was the right thing to do. If you hadn't been out of the room having a crafty fag, you would have heard the rest of that experience. "Hell is in the heart of a man who hates his father", she quoted from Nichiren's writings, and she'd been hating her father for over 20 years. But, in the course of a month's chanting she came to realise that the only person suffering from that hatred was herself. Her father was quite unaware of her suffering. And when she commenced dialogue with him and learned his point of view she stopped wanting him to die. They are now learning to be friends.'

'Fair enough,' I said, suitably chastened. 'But...and I'm really sorry to bring this up again...what about that helicopter?'

Yes, I really was that shallow in those days. Having grown up in near poverty, I was pretty much obsessed with making money. My whole childhood had been spent trading rare pennies and then stamps with geeks in mackintoshes. Then I had spent my whole adolescence and most of my early youth chasing down the best-paid jobs, even though I had little or no ability to do any of them. Now, I was hoping, Buddhism would be the fast track to lots of lolly. I could have whatever I chanted for, right?

'Well, not quite,' said Nick when I posed the question. 'It's not just a matter of chanting for things. You have to take some action as well.'

I asked him what kind of action, and he put me to work making little plasticine pigs.

'There you go,' he said, showing me the thriving cottage industry going on in his garden shed. 'I wholesale thousands of these little critters out to big shops. That's how I got the helicopter.'

I was rubbish at making little plasticine pigs. Their ears kept dropping off and the more I tried to make them stand up, the more they fell over. I tried to tell Nick I wasn't gifted in the plasticine pig making department, but he was not sympathetic.

'You haven't got much patience, have you, Frank?' he said. 'Keep at it!'

'Those pigs don't like me,' I told him miserably. 'What else have you got?'

Nick grinned. 'I've got cats. You like cats, don't you? Yes, why don't you try cats instead?'

What I did to those poor little plasticine pussies was an abomination. One of them stared accusingly at me through two horribly mismatched eyes while another fell off the table and raised its twisted paws up from the floor in an attitude of mute supplication. 'Put us out of our misery!' it seemed to cry. 'Don't maim and disfigure us anymore!'

But then, on the 8th of May 1983, I got my Gohonzon and plasticine pets were the least of my worries.

My whole life was about to be turned upside down.

Chapter 11

Devils and Bingo Balls

I suppose, in the end, my decision to become a bona fide Buddhist owed as much to surviving six months in the most gruelling job I had ever had than to any external pressure from other Buddhists. Old Bill, Bertie, Matron, Mr Parker, I would have fled from them all had it not been for the calming influence of this weird and wonderful chanting that threw up solutions whereas previously I would only have seen problems. Yes, it was time for me to stop ducking and diving around life, and to commit myself to something that confronted it head on. No longer could I only keep myself going at the old people's home by telling myself: 'This isn't forever. I can get out any time I want.' Now, for better or worse, I was in for the long haul.

But if I had abandoned my flippant attitude towards Buddhism and had finally come to take it seriously, I had not counted upon the reaction of my hard-line Catholic mother.

'What have you got against Jesus?' she wept when I told her the good news of my conversion. 'He died to redeem our sins. How can you abandon Him?

'I haven't got a problem with Jesus,' I replied, still scarred from my school days. 'Just all those priests He keeps employing!'

A feeling of excitement, mixed with more than a twinge of trepidation, seized me as the bald Japanese priest bowed and handed me my small devotional scroll in the Richmond Centre. 'Wow, I've got my very own gohonzon,' I thought to myself. 'Now, what do I do with it?'

The first thing I did with it was arrange a small enshrinement ceremony back at my flat in Clapham. Brenda was there, and so was her sister Anna, and so was Nick, despite my murderous designs on his garden shed animals. As incense was lit, and my tiny bell tinnily intoned in the background, Anna removed my scroll from its envelope, carefully unrolled it and hung it even more carefully on a hook inside the cheap wooden altar or *butsudan* I had purchased earlier. 'Congratulations, and good to see you again my long time karmic friend,' wrote Brenda on the envelope. 'Continue chanting no matter what,' added Anna under that. 'Remember, winter never fails to turn into spring.'

Then, when the crisp white scroll with the bold black lettering was correctly positioned, we all chanted for a bit and closed the door on it.

I was officially a Buddhist.

Three days later, I had reason to be very thankful for that. Without warning, in the middle of the morning, the door to my office in the home flew open and Mr Parker, accompanied by Mr French, the Committee Treasurer, charged in.

'Don't say anything,' muttered the red-faced Chairman, jabbing a key into the safe and pulling out the cash box. 'Don't say a word.'

My mind went into free fall. What on earth was happening here? Had I done something wrong?

Too Young to be Old

I watched in dumb silence as Mr Parker flicked through all the pocket money envelopes I had earlier collected from the post office. Then I saw him pause and pick out one of them.

'So it's true,' he murmured as he opened it up. 'Been nicking money off my mum, have you, Mr Kusy? How long has this been going on?'

To say I was flabbergasted would be an understatement. I stared at Mrs Duff's pocket money envelope – turned upside down and devoid of contents – in total disbelief.

'I...I don't understand,' I stuttered helplessly. 'I'm sure the money was there when I put it in the safe.'

'Well, it's not there now,' said the grim-faced Chairman. 'No, I've got you bang to rights, Mr Kusy. You're going down for this.'

And before I could stop him, he marched over and began frisking me, digging his rough, dirty-nailed fingers into my pockets one by one, then patting me down as though I was some kind of drug smuggler. In the background I could see timid Mr French shrugging as if to say 'I'm sorry, but what can I do?'

As shock gave way to embarrassment, I pushed my assailant away. 'Get *off* of me!' I cried unhappily. 'I haven't *got* your money. There's been a mistake!'

'Oh, there's been a mistake alright,' puffed Mr Parker. 'And you're the one who made it. So the cash isn't on you, have you spent it already? I see you got a new car!'

Light began to dawn. I had given up on the helicopter idea for the time being, but I had been chanting to afford a car. It would be nice, I reasoned, to run my mum down to the coast at weekends, or simply to do errands from the home that were taking up a lot of time on foot. And then Greg, another Buddhist friend of mine, had rung and offered me his car – a very nice Ford Cortina – for next to nothing because he was moving to Spain. 'What luck!' I had thought at the time and drove it straight to the home to show it off to everybody. But I had been surprised by their reaction. Only John Gray had showed any interest – all the rest of the staff just looked sullen and resentful.

'Of course, how stupid of me,' I thought, giving myself a mental slap. 'None of them will ever have money to buy a car. I've made enemies here. And one of them has set me up for a fall. Now, who could it be?'

'I got the car cheap from a friend!' I protested to the Chairman as my mind began feverishly ticking off potential candidates. 'I did *not* steal money from your mother to fund it!'

Mr Parker had that look on his face again. The conflicted one. Part of him, I could see, wanted to believe me. But the greater part just couldn't.

'Okay, Mr Kusy,' he said slowly. 'I'll do you a deal. I'm going to walk out this office now, and I'm going to walk back in again in ten minutes. If the money's back by then, I'll just accept your resignation. If it's not, well, it's the police station for you, young feller me lad. You just made my shit list. Congratulations.'

I clung, white knuckled with rage, to my desk as he made his exit. Then I ran out of the home and banged on the bonnet of my new car so hard that I nearly broke my wrist. 'Who hates me this much to do this to me?' I wept in frustration. 'Who would be so mean?'

I raised my numb hand to my face and as I did so, I heard a faint sound to my left. And looking through the thin pane of ground floor window glass that separated us I saw Mr Bragg. He was laughing at me.

They say it's better the Devil you know, rather than the one you don't, but it didn't feel like it. This particular Devil was holding all the cards. Wiping the tears from my eyes, and trying to fight down the urge to commit some very un-Buddhist violence, I strode back into the home and started to chant. And from the depths of my subconscious, as I began to master the turmoil of my tortured mind, an idea came up that was so simple and so cunning that I began to smile.

'Can I have a quick word with you?' I summoned the smug-faced handyman. 'In private.'

I waited till he had sat down in my office, and then closed the door.

'Why?' I said quietly. 'And how?'

Mr Bragg's eyes darted left and right. He was looking for a trap. Then, having satisfied himself that nobody could possibly be in earshot, he went off on his rant.

'You think you're "it", don't you, laddy? Before you came, I had everybody – Matron, Parker, the whole Committee wrapped round my little finger. I knew Douglas Bader, I did. Yes, Dougie and me went way back. I was with him when he was sent to Colditz and told everyone he'd be "a plain, bloody nuisance to the Germans". Now, I'm being a plain, bloody nuisance to you! You didn't know I'd waxed all your keys, did you? Or that I'd have time to get in your safe and get Parker's silly old mother's cash while you were showing someone else your stupid car. And the best thing is, there's nothing you can do about it. Aye, laddy, your days here are well and truly over.'

I looked at my watch. The ten minutes were nearly up.

'Yes, it does look like it,' I said with a sigh. 'The Chairman will be back in a few seconds, if you want to gloat.'

Mr Bragg's eyes widened with anticipation. Here was an unexpected bonus, a ringside seat to my humiliation. It was more than he could have hoped for.

Moments later, bang on cue, Mr Parker and Mr French swept in. They looked curiously at the grinning, expectant handyman, and then they looked at me.

'Okay, Mr Kusy,' said the scowling Chairman. 'Which is to be: the cash or the cop shop?'

I waited a moment to let the excited thump in my chest die down. 'Neither, actually. I think you might like to hear this.'

Three pairs of eyes trained down to my waist as I withdrew the Sony Walkman from my pocket. And then Mr Bragg gasped as he guessed what was coming next.

'...Before you came,' intoned my trusty recording device, *'I had everybody – Matron, Parker, the whole Committee wrapped round my little finger...'*

The silence when the Walkman clicked off was deafening. A pin could have dropped in Calais and we would have heard it.

Then the Chairman rounded on Mr Bragg.

'Wrapped me round your finger, is it?' he said with a snarl. 'Who the fuck do you think you are, you jumped up excuse for a caretaker? I bent over backwards to get you your buggering contracts of employment – yes, I know it was you behind that – and this is how you repay me? By stealing my "silly" old mum's money and making a fool of me with Mr Kusy? You cum 'ere, I'm going to tear you a new one!'

Mr Bragg may have been pals with WWII flying-ace-with-no-legs Douglas Bader. He may also have faced down the Germans in Colditz. But nothing could have prepared him for the human pit bull that was Mr Parker. With a short grunt of rage, the Chairman lifted the hapless handyman off the ground by his boiler suit lapels, banged his head repeatedly against the wall, then frogmarched him into the street and deposited him on the pavement with a well-directed kick to the posterior.

'I should report you to the police,' he said, dusting off his hands. 'But you're not worth it. Scum is scum, and you're the worst kind. Now, sod off out of my home and never come back!'

*

'How do you fancy helping me run the Bingo?' said John Gray as he came upon me slumped on my desk. 'I've just heard what happened. You need a distraction.'

My head was not exactly in the right place for bingo. An hour had gone by since the culprit had been caught and my reputation had been restored. But even though Mr Parker had shaken my hand and apologised profusely, I was still depressed by the sheer unfairness of the incident. After all I had tried to do for the home – I was now slaving away up to 60 hours a week – how could he even have suspected me?

But maybe John was right. Maybe I did need a distraction.

'I've never played before,' I mumbled through one half-opened eye. 'Is it easy?'

'Well, it'll certainly snap you out of your trance,' grinned John. 'I'll meet you in the leisure room at six. Do *not* be late.'

I dutifully turned up at six sharp and saw why promptness had been of the essence. At one second past six, John opened the door to the leisure room and about thirty residents and friends piled in.

'Hang on to your hat, Frank,' John told me. 'They take their bingo very seriously.'

How seriously they took it terrified me. 'Get out of my seat!' ordered an hysterical Mrs Caitlin. 'That's my lucky seat!'

'Sit behind me, Frank,' laughed John. There's always a bit of a scrum to be in a good position to the caller.

'Blimey,' I said, backing away from the oncoming tide. 'Are they all this superstitious?'

John's grin widened. 'You wouldn't credit it. Last year, a woman turned up with an urn with her brother's ashes in it for "luck". We only turned her away with difficulty.'

As a bingo virgin, I had no idea of what to expect. All that I knew was that I had been given the job of selling the bingo cards. I was surprised to find most of the old folk wanted ten. Ten! How were they going to keep track of ten cards? Then I noticed the specialist equipment they had brought along...their brightly coloured bingo pens and the very large bags for taking home prizes.

I only had a pencil and had purchased just one card in my brave attempt to join in. 'How hard can it be?' I thought smugly. 'All these old ladies who play, surely my reactions are quicker than theirs? Besides, it's just for fun, isn't it?' How innocent I was, and how terribly wrong. That first evening I was like a child walking barefoot into a nest of vipers.

Too Young to be Old

As soon as John called 'eyes down' silence fell over the room as patrons dabbed for a potential win. Heads lowered in fierce concentration, they looked like woolly-jumpered horses at the starting gate. Then John pulled the first numbered ball from the large fish bowl in front of him and they were off!

'One little duck....number 2,' he called, and half the competitors started quacking.

'Burlington Bertie...number thirty!' he said, and all the old ladies looked enviously at Betsy.

'Eighty-Three...stop farting!' A chorus of 'Who, me?' ran round the room.

I was finding this alarming. This bingo thing seemed to have a language all of its own. But the main thing that shocked me was how quick the old ladies were. They checked over their ten cards while I was still searching through my single one. I felt so stupid. Then there was the difference between a line and a 'full house'. It sounded easy, but in the adrenaline rich atmosphere of the competition, I was finding it hard to remember if we were going for a full house or a line – something I found to my cost two minutes later.

'I've got a line!' I shouted in a nervous voice, and John stopped calling.

Thirty pairs of rheumy eyes tracked in on me. Then the mutterings began. I had dared stop the proceedings for a line when everybody else was going for a full house!

'So sorry, everybody,' John apologised for me as I went bright red. 'It's Mr Kusy's first time. Let's push on!'

After that, of course, I didn't dare call out anything. But I was not the main offender that afternoon. One old lady called 'House' too early, an unforgivable sin, and all the others shifted their chairs and turned their backs on her.

Yes, this bingo was a very serious business. But John had

been right – I got so caught up in the buzz and excitement of it, I forgot all about my rude treatment earlier on. Indeed, by the time I'd finished handing out the prizes, basking in the nods and thanks of the lucky winners, I was wearing a big smile on my face.

A smile that was about to get a whole lot bigger.

Chapter 12

Anna

As life in the home settled down and no more dramas were in the offing, my mind began to turn to romance. My relationship status had been single for close on two years now, and notwithstanding my brush with the teddy bear hugging nymphomaniac, I thought I might be ready for love again.

'If even Bertie and Betsy can get it together,' I told myself, 'why not me?'

I started my search at a dating club called London Link Up, which held regular parties at a venue near Leicester Square. But this didn't work out well. One night, I hooked up with a girl called Mary, who wore a lot of leather. I couldn't remember what happened between us, I was so drunk, but I certainly did remember what happened the next morning. I walked into her kitchen, in search of a glass of water, and surprised her in the act of making a telephone appointment with a "client". 'Are you a prostitute, then?' I asked her, and she said: 'What are you complaining about? You're getting it for free.'

I told Brenda about this and she just laughed.

'If you want to change your relationship karma,' she said, 'you should become a group leader.'

I flinched. There was that word 'should' again.

'What?' I said, a look of horror on my face. 'As in a Buddhist group leader?'

'That's right, silly,' Brenda said with a giggle. 'It will develop your caring, nurturing side. You do have one, don't you?'

Yes, I did have a caring, nurturing side, but I had to confess, only when it suited me. I was a typical only child, everything on my terms. And the thought of having to look after a group of other Buddhists filled me with dread.

'Look, I have to be honest,' I said. 'I've only just got used to the idea of managing the old people's home. I really don't think I'm ready for any more responsibility.'

Brenda's eyes crinkled with amusement. 'That's what Dick Causton said when President Ikeda asked him to take charge of our Buddhist organisation for the whole of the U.K. And Ikeda came back with just three little words: "Responsibility needn't weigh."'

'Yes, well,' I replied. 'From what I've heard, Causton was an army colonel before he became a Buddhist. He was used to man management. Me, I'm hopeless. I can't even run the home on my own. I've basically delegated all the man management to my deputy, John Gray.'

But my protests fell on deaf ears. Brenda got together with her sister Anna, and a week or so later I found myself appointed leader of 16 Buddhist members in the Clapham/Balham area. I looked at the list of their names in dismay. I didn't know any of them. I wasn't sure I *wanted* to know any of them. All that I did know was that they were going to be a drag on my precious time. 'Blow this for a lark,' I thought to myself. 'I'm knackered as it is, coming home every night after ten hours of work. I can't be arsed to make 16 new friends.'

Too Young to be Old

But then the monthly discussion meeting came around and although I hoovered my little flat and tried to make it look as welcoming as possible, only four people turned up.

Brenda was not impressed when she heard the news. 'The monthly discussion meetings are the most important dates on our calendar,' she scolded me. 'Did you even bother to pick up the phone to anyone?'

'No, I didn't,' I replied sulkily. 'I've got a thing about telephones. I've also got a thing about talking to total strangers.'

'Well, they wouldn't be total strangers if you talked to them!' sniffed Brenda. 'Honestly, Frank, for a bright bloke you can be awfully dim at times. How can you expect to change your relationship karma if you can't be bothered to make any new friends?'

She had a point of course. All my life, I had hidden away from people, much preferring my own company. Even as a child, if the occasional school chum turned up at my door asking if I wanted to come out and play, I would shout down the stairs to my mum: 'What do they want to do?' And if it wasn't something I wanted to do, I would just leave them standing there while I returned the book or comic I was reading.

Now, as I approached my 28th birthday, I realised the need for change.

'I've got to get out of my comfort zone,' I told Brenda's sister, Anna. 'I haven't had a relationship in two years.'

'Neither have I', said Anna glumly. 'I know. We'll do a *tozo*.'

'A *tozo*? What's that?'

That's when, if you have a particular problem to deal with, or you want to shift a particularly large chunk of karma, you chant for a long time.'

'How long a time?'

'Oh, well, at least three hours.'

Three hours? I had never sat still for three hours in my life before. Even a visit to the cinema had me checking my watch every ten minutes to get out.

But it was strange. After the first half hour, with seven other Buddhists packed into my tiny flat a few days later, my mind settled down and stopped running around like a restless rabbit. I began to zone in on the heart shaped character of *myo* in the centre of the gohonzon. And then, without being even conscious of it, I began to reflect on my relationships.

What did the women in my life so far have in common? Well, for one thing they all were more than slightly unhinged: from the predatory Liz who seduced me at university and then kicked me out of bed when she found I was a virgin, to the sweet but silly Lucy who liked baby talk and who had an annoying laugh, to the cold and hysterical Barbara who stopped her car on the fast lane of the M25 motorway and demanded a £900 engagement ring, and of course to the sex mad Christine who dumped me in favour of someone with a more proactive Percy. Not to mention the bubbly fatling Sue whose father was a famous film director and who flew in with his helicopter every weekend with bags of sweets and treats for his favourite daughter.

And herein lay the second common denominator. All of these women had serious daddy issues. They were all prima donnas who had been spoiled rotten when young and who still wanted, no, *expected* to be looked after by someone else. I hadn't felt half of a couple at all.

As the tozo entered its third hour and I achieved a state of total concentration, I was able to put my 'self' aside completely and feel 'all that' out there in the big, wide world and not be distracted by anything. I lost all sense of time. And at the end, all I could see was just the characters on the gohonzon, and a sort of gold blur in the background. I wasn't aware of thoughts *per se* at all.

But one thing I was aware of. My relationships with women had been based on fun and sex, and that simply wasn't enough. What I needed now was someone in my life who would be a rock, who would be a real friend and a true 'equal' partner, not just in this existence but in all existences to come.

And who did I get?

I got Anna.

The tozo group moved straight on to a party at a mutual friend's place. The mutual friend was called Carol and she had just got together with another mutual friend, Graham. And it was here, during a particularly vigorous jive dance with Anna that we lost a grip of each other and went flying into Carol's record player. I don't remember much of what came after, only that the record player cut a deep groove into my arm and Anna drove me straight back to her place and put me to bed and started dressing it.

Then, as she finished administering to my wound, something happened. There was a moment's pause as she leaned forward to inspect the bandage, and in that moment – as our eyes met and the heat unexpectedly rose between us – I drew her down and kissed her.

'What do you think you are doing, Mr Frank Kusy?' she murmured huskily.

No more words were spoken.

I woke the next morning feeling more relaxed and content than at any time in my life. The sun was streaming in through the window, bathing the tidy pinewood dressers and wardrobes – and even tidier collage of blue and white plates on the wall – in a warm, rosy glow. Downstairs, I could smell eggs cooking.

My stomach gurgled. I hoped some of those eggs were for me.

'My arm's feeling a lot better,' I said as I drifted down the stairs in one of Anna's much-too-small dressing gowns.

'I'm guessing the rest of you feels a lot better too,' she replied, an amused smile playing across her small, pretty features. 'Do you mind telling me what happened back there?'

It was a good question, and one for which I had no easy answer. We had been friends for three years and in all of that time neither of us had evinced the least romantic interest in each other. In fact, we had had trouble even being friends. I found Anna too serious – my humour fell flat on its face with her – and all through our days at the Financial Times, where she had been my direct line editor, she had found me too 'bouncy'. Imagine our surprise when we found that we were perfectly suited in the bedroom department. Seriously bouncy, indeed.

'I don't know,' I ventured hesitantly. 'But I tell you what, those *tozos* work a treat. When can we do another one?'

Anna's dimpled smile deepened. 'Let's take it slow, sweetie,' she replied. 'But get this. I've been on the phone for the last hour, and it turns out that all eight of us who did that *tozo* got off with someone at that party. In fact, we all got off with each other!

'Even John, that gay guy?'

'Yes, even him. Who knew that Stewart, that new member who turned up late, was gay too?'

Anna's house was actually half a house, the top half, and it was beautifully decorated and very, very clean. She also had two cats – Bitsa, a very affectionate tortoiseshell and Isis, a rather more aloof black cat. For someone like me, who loved cats but had never had one of my own, it was paradise. I stroked and fussed over those cats like nobody's business and

Too Young to be Old

Anna, bless her, never got jealous. She understood a man with cat deprivation issues.

I suppose it was when I started doing a monthly magazine for the old people's home that Anna took an interest in my work. Her skill set was print and design and she was using it to good effect for a well-known charity in central London. 'Okay, I'll help you with your magazine,' she said one day. 'But first, let me see what we're dealing with – I'd like to visit the home.'

Looking back on it, she may have regretted that decision. The day I chose for her visit was the home's annual Garden Fete, and no sooner had we arrived than Old Bill kicked off in best belligerent form.

'I'm not drinking this,' the truculent troublemaker was howling down the corridor. 'It's got bromide in it!'

The object of his derision was a cup of tea, which he was

brandishing in the air like an Olympic torch. And as soon as he caught my eye, he advanced on me, his hands shaking and the tea slopping all over the floor.

'What's the story here?' I hurriedly enquired of Matron, who had just turned up. 'And what's bromide?'

'They used it during the war, dear,' said the Matron. 'For the soldiers in the army, to control their sexual appetites. That's what's got Bill going. He's sure we're putting bromide in his tea.'

'I know it,' proclaimed Bill, coming to a stop about three feet away from us. 'I can taste it!'

Matron eyed the tea-slopping rebel dubiously, and decided she'd had enough.

'Now, come along, dear,' she told him. 'If you've got bromide, so has everybody else, because everybody's tea is poured from the same pot!'

'Well, there you are, then!' crowed Bill triumphantly. 'They've all got bromide! But you know what that lot are. They won't talk. They wouldn't even notice the difference. But I do. And I'm not having it!'

Matron didn't see the white stick coming. It whistled past my ear and into her perfect pile of tightly permed hair with a loud *thonk!* As she let go a light moan and fell to the floor, I looked across nervously to Anna. How was I going to protect her from the craziest old codger in Clapham?

But she didn't need my protection. With Bill still raging: 'Yes, as soon as my gammy leg's better, I'm going to take this cup of tea over to the chemist and have it analysed for bromide!' she stepped forward and gave him a big hug.

Everything stopped. Bill stopped ranting and looked over her shoulder with a look of surprise. Matron stopped scrabbling around on the floor and regarded Anna with an even bigger look of surprise. And I stopped reconsidering my career in elderly care and wondered why none of us had thought to do this with Bill before.

'He just needs a bit of attention,' Anna whispered to me. 'That's all he wants.'

*

Out in the garden, the fete was in full swing. Mr Parker was manning the book stall, Miss Sherring was handling the guess-the-weight-of-the-cake stand, and over in the corner Betsy and Bertie were running the tombola.

I wasn't sure where to take Anna first, but then Bertie spotted me and summoned us over urgently.

'Oi, Mr Kusy, come over here! Me and Betsy's getting married. I just called the Pope in Rome, and he's flying over to do the honours.'

'Pope John Paul II?'

'Yerse, that's the one. He's on his way over right now. Me and Woj – that's Wojtyła to you – go way back. He was a pretty nifty footballer in his day. I used to peel his oranges for him at half time.'

Anna and I shared a secret smirk. She'd heard all about Bertie and his fanciful imaginings.

'Is that your young lady?' intervened little Betsy, peering at Anna through her dark glasses. 'My, but she's a pretty one!'

'Yes, she is,' I said proudly. 'And she's pretty smart too. She just shut down Bill in the middle of one of his rages.'

'Hmphh,' muttered Bertie. 'What that Bill needs is a tot of rum. That'd make a man of him.'

'I hope you don't mind me saying this,' said Betsy, grabbing one of Anna's hands. 'But if was 30 years younger, I'd make a man out of Mr Kusy. It took me ten minutes to do up the buttons on my blouse this morning. He should have been there!'

Anna laughed and sat herself down next to my aged groupie. 'No, of course I don't mind. But tell me, Betsy – I've heard so much about you – how did you feel when you first came into the home?'

Betsy went suddenly serious. 'Oh, very distressed at first. I had to leave my pussy behind. I missed him terribly. And then they put me in the TV room, bang next to the television. I don't like television, never have. But this was the only seat they had left when I came in. I had to wait until someone died before I got another one.'

'Oh, that must have been awful,' sympathised Anna. 'I'm not a big fan of television either. It kills the art of conversation, doesn't it?'

Betsy nodded in furious agreement. 'When I was young, we used to play tennis or go swimming. Or read, or go to dances, or see friends. We were never bored. Nowadays, everyone *rushes* to the station. Even the young ones. And they *rush* home, and they sit *down!* And if there's nothing on the telly, they sit around and wait until there *is* something on the telly. I can't understand it. They've got the whole of London to explore, and they can't be bothered to walk around it. The best thing that could happen to this country would be if there was no television for six months! It's a drug, a terrible drug.'

I opened my mouth to comment, but was silenced by a booming voice on a megaphone.

'Okay, you lot,' Mr Parker addressed us. 'I got a special announcement. My mum, Mrs Duff, has her 100th birthday today. Yes, that's right, she finally made it. So let's all gather around and give her a big cheer!'

There was a moment's silence as the Chairman's white-haired old mother was wheeled onto the lawn, and then everybody launched into three choruses of: 'Hip, hip, hooray!'

Mrs Duff looked happy, but confused. And her confusion deepened when Mr Parker handed her the congratulatory card from the Queen. She studied the picture of our most gracious Majesty, and then the gold embossed message inside. Then, as we all stood around her in expectant silence, she lifted her tired old head and fixed us with a look of puzzlement.

'Elizabeth? Elizabeth?' she enquired querulously. 'Who is

this woman? I don't know her!'

*

'So,' I said to Anna as we walked back to my flat at the end of the day. 'Do you still feel like helping out on the home's magazine?'

'Of course,' she said with a smile. 'But honestly, Frank, a magazine is not enough on that place. You should write a book.'

Chapter 13

The Importance of Mission

As 1983 slid into 1984, I had reasons to be cheerful. Not only was my relationship with Anna going well – we were talking about a holiday together on the island of Skiathos in Greece – but I had finally got my head around being a Buddhist group leader. Yes, the lash of Brenda's ridicule had left its mark. Instead of just staring at the list of local members' names I had overcome my painful shyness and had started ringing up names on it. Result? At the second meeting at my house all 16 people turned up. Plus two guests and a dog.

Just one last Buddhist-related hurdle remained, and it was one I had been putting off for a long time. I was going to have to meet Richard Causton.

I was not looking forward to this. If Nick, my helicopter owning friend, had problems with authority figures – in particular with male ones – they were as nothing compared to mine. Okay, I had my grandfather to look back on, he was a male role model to treasure, but I still – at coming up to 30 years of age – had no-one I could look up *to*. I hoped against hope that Dick Causton could be that someone. If he wasn't, if the big chief of our small Nichiren movement in the U.K.

turned out to be a disappointment, I wasn't sure I could handle it.

But I needn't have worried. Tall, dignified and charismatic, Dick impressed me from the start. I was particularly impressed when, at my first big leaders' meeting at Euston House, he spotted two guys asleep at the rear of the hall. 'WAKE UP back there!' he thundered, bringing down his fist on a table and nearly smashing it in half. 'You can sleep long enough when you're dead! We're supposed to be roaring lions for world peace – WAKE UP to your mission in life. Wake up NOW!'

Mission? What was he talking about? I was honouring the memory of my grandfather by protecting the rights and dignity of the old people in my care. Wasn't that enough?

I didn't get a chance to ask Dick this question at that big meeting – there were over 100 people present and, to be honest, I had been a little intimidated by his show of temper – but then, a few weeks later, he unexpectedly accepted an invitation to attend a much smaller meeting at Anna's house in Crystal Palace.

'Nichiren Daishonin was the son of poor fisher folk,' he told us when I finally got a chance to speak up. 'He spent his whole life battling arrogant priests and evil authorities to reveal his mission, which was to return the promise of the Buddha – happiness in this world and peace and security in the next – to common, ordinary people just like him. We have got to do the same as Nichiren; we have got to find our mission. We have got to discover who we really are, what our bigger selves are, what we're here on this Earth to achieve. And then, having discovered this, to go ahead and fulfil it. With the fearlessness and the joy of Nichiren Daishonin.'

I blinked. This mission thing was obviously big on Dick's agenda. But how did that relate to me?

Too Young to be Old

'Chant to see your *kyo*', said Dick when I posed this second question. '*Kyo* is the final syllable of Nam myoho renge kyo, and it means "sound and vibration". Every living thing has its natural path in life – the way it vibrates most naturally with its environment. Fish swim naturally, birds fly naturally, but human beings, blinded by illusion and the three poisons of greed, anger and stupidity all too often miss their natural calling in life. I would like you – not just you, Frank, but all of you – to go away and chant to see your *kyo*, to see where you can become your greater selves and create maximum value for *kosen rufu* or world peace.'

Dick's words made a great impression on me. I went straight home and chanted five hours to see my *kyo*. Then I booked myself onto a Buddhist course at Trets in France and chanted for a whole week to see my *kyo*. And at the end of all this, when I came to have personal guidance with Dick, I told him I wanted to become a writer. It was a dream I had had when a child, but one that had been drummed out of me by my

'Get a proper job' mother and by the Jesuits.

'That's jolly good timing,' said Dick. 'I need a writerly type right now. I'm thinking of putting all my lectures on the basics of Buddhism into a book called *Buddhism of the Sun*. Have you been to any of those lectures?'

Well, yes, I had been to all those lectures. Even the one that had collapsed in hilarity when Dick's Japanese aide had leapt up at the end to say: 'T'ank you, Mr Cawstun. Now, evelybody *crap!*'

'I even taped them,' I told Dick, showing him my trusty Walkman. 'So you can have an oral transcript to go with your written notes, if you like.'

Dick beamed. 'That would be great. Though you'll have to work with the primary editor, Jim Cowen. Will that be a problem?'

'No problem at all,' I said breezily. 'I'm good at working with other people.'

I was deluding myself. I was not good at working with other people at all. Especially when the other person turned out to be the most arrogant, annoying know-it-all I had ever met. Jim was a brash, red-headed powerhouse of energy with the word 'ego' stamped all over him. I tried to like him, but just couldn't – everything I suggested to Jim, he had a contrary (and better) opinion about, and every new idea I came up with, he promptly adopted as his own.

The crunch came when, after two weeks of laborious banging away on my typewriter, I produced transcripts of all six of Dick's public lectures.

'Oh no,' sniffed Jim, flicking dismissively through them. 'We can't be doing with *those*. You've even put "loud laughter" in brackets after what you perceive to be the funny bits. Are you trying to trivialise the serious nature of Buddhism?'

'No, I'm not,' I replied hotly. 'People *did* laugh at those lectures. Dick used humour a lot to get his message across. And of course we've got to use these transcripts. That was specifically why I was brought into the project!'

Dick walked in just as we were about to come to blows. 'You are two of the angriest people I've ever come across,' he said with an amused expression on his face. 'I feel like banging your heads together!'

We looked at him in puzzlement, especially me.

Moi? Angry? What was he talking about?

'He's talking about your fundamental life-state, Frank,' laughed Anna when I told her my experience. 'Of the ten "worlds" of Buddhism, from Buddhahood and Bodhissatva to Hunger and Hell, you exist mostly in the world of Anger. And don't tell me I'm wrong – I've worked with you myself at the F.T., remember, and you couldn't be "told" anything by anybody. You were a right stroppy grump most of the time.'

I didn't appreciate what Anna was saying, I had always considered myself a quiet, easy-going sort, not given to anger or violence at all. But then my mind fled back to the day I'd finally got fed up of being picked on at school and had beaten up a bully at a bus stop. And then to the day I had hospitalised my two room-mates at university for playing loud music when I was mugging up for my finals. A white heat had descended upon me on both occasions – I had become anger and violence personified.

'So what do I do about Jim?' I said, deflecting Anna's observations.

'If I know Jim', she replied, 'he's probably resenting someone muscling in on "his" book. And you're probably pushing your ideas a little too forcefully. The positive side of anger is courage, sometimes the courage to let go of one's pride or ego.

Try praising or encouraging him a bit, even if you don't mean it. Keep your final goal – the best possible book for Dick – in mind. That's worth a little give and take, isn't it?'

I decided that Anna must be right and to my surprise, three months later, *Buddhism of the Sun* – the UK's first home produced book about Nichiren Buddhism – went into print. Even more surprising, by this point, Jim and I had become friends. We were never going to see eye to eye on everything, that was clear, but both us had gained enough trust and respect for each other to bring a common objective to completion.

And to change quite a big chunk of karma in the process.

*

A month or so later, I returned to the home after a lovely holiday in Skiathos with Anna to find Old Bill standing furiously in the corridor.

'I'm not going!' he ranted, his white stick flailing about in the air. 'I'm not going, and you can't make me!'

'What's the story here?' I accosted John Gray. 'Are you taking him somewhere?'

John ran a hand through his thinning brown hair. 'It's a black day for Bill,' he said. 'He's going to hospital.'

'Hospital? He looks alright to me.'

'Yes, I know he does,' said my phlegmatic deputy. 'But the drugs he takes for his depression, they make his hands shake. He can't hold things anymore and shouts for help to fasten his trouser fly buttons. You missed it, Frank, all that shouting has really got the other residents down. This morning he was complaining of "shortness of breath". He demanded oxygen. He probably used all his own up shouting. He's terrified of going to hospital, thinks he's never coming back, but he's got to go and there's an end to it.'

I looked over John's shoulder and spied Bill upping his protest. The decorators were in, papering the reception lounge, and Bill was tempting fate by walking back and forth under their ladders. Mrs Lowry, a dotty old lady with an armful of toilet paper she had just removed from the toilet, tried to remonstrate with him but was instantly attacked. Bill seemed to want that toilet paper, I couldn't imagine why.

'Oh, he's getting quite impossible!' fumed Matron, turning up with a wheelchair. And before Bill knew it, he had been shoved into it and out into the waiting ambulance.

All was quiet for the next three days. Then, on the Friday, Old Bill returned. He wore a look of betrayal and scowled at all the staff.

'I have suffered a terrible shock!' he proclaimed, and retired to bed.

'He should never have come back,' scowled Matron. 'I don't understand that hospital. Why do they keep sending him back?'

I had mixed feelings on the matter. Yes, Bill was a pain, but the home had been dull as ditchwater without him. No fun at all. Part of me was really looking forward to him being back in action – tormenting the staff and demanding their respect – when I next turned up for work on Monday.

But it was not to be. Back at the home after a quiet weekend with Anna, I detected a tangible air of gloom. It hung over everybody like a shroud.

'Something's happened, hasn't it?' I enquired of Miss McCann, who was lingering outside my office.

The sallow-faced deputy Matron drew me close. 'It's Bill,' she said quietly. 'He's gone, I'm afraid. He went last night.'

'Gone?' I said in shock. 'What do you mean, "gone"?'

Miss McCann's expression remained detached. 'Saturday, he refused to get up. He said he had a "frozen arm". He'd got a chill on it, he said, from opening the door for Matron. He wanted to be spoon fed by the staff, 'cos he couldn't eat with a frozen arm. Well, we thought he was just playing up again, and stuck him in the patio with his breakfast, expecting him to get on with it himself. But no, two hours later, he's dead! He really *did* have a frozen arm. It was a stroke, and the arm was the first thing to be paralysed. Poor old Bill. He kept telling me, that morning, about a "visitation in white" he'd seen in his sleep. I thought he was kidding, but maybe he did see the writing on the wall…'

A deathly calm settled over the home that Monday. All the residents knew Bill was dead and none of them wanted to talk about it. But most of them turned up for the funeral. Matron said it was very necessary for them to be allowed to attend. Painful perhaps, but very necessary. The final stage, as it were, on which each one of them would give their final performance. And much better than just telling them: 'Oh, Bill has gone to heaven.' Elderly people were all too prone to ask awkward questions like: 'When will he be *back?*' Matron didn't see why they should be shielded from that. It was then, and only then, that I realised that she cared.

Old Bill used to like singing first World War songs like 'It's a long, long way to Tipperary' and 'Hang out your washing on the Siegfried Line.' He was a fat, bald man of 82 and no sooner had he gone than I realised how much I was going to miss him. He was always coming up and asking: 'Am I alright?' And I'd say: 'Yes, Bill, you're fine.' And he'd smile and say: 'Oh, that's good,' and be no trouble for the rest of the day. All he wanted out of life was reassurance and attention. 'Rather too much attention!' grumbled little Betsy when the subject came up. 'But I cried when I knew he'd gone. He was *ever* such a fat man, wasn't he? He used to say Mr Parker was his father. How *could* he be?'

Chapter 14

Too Young to be Old

The passing of Old Bill marked the end of an era. Not just for me personally – all of a sudden, my 'mission' in Clapham didn't seem so important anymore – but for the home in general. A cluster of other residents followed Bill in quick succession, and in their place came a cluster of Maggie Pratt style dementia and incontinence cases.

'Trouble is,' I observed to John Gray, 'the elderly sector is changing so fast. A year or so ago, as you remember, the home was in danger of closing because the committee was reluctant to admit infirm or mentally disabled people. Yet that was all we were being offered. Now we have twice as many wheelchair patients and cases of senility as before, staff have had to receive nursing training, and a whole new set of residents have arrived.'

John nodded in sad agreement. 'Yes, it's just as Mrs Teasdale predicted. We've become nothing more or less than a nursing home. All the elderly with some mobility and wits about them are being kept in charge of their relatives – simply because, unless the relatives can afford to send them to expensive private homes, the Social and the Council won't subsidise their fees. They will now sponsor only elderly people with real physical or mental debility.'

Some of the new breed of residents were not a problem. Dotty old Miss Gillings, for instance, was rather sweet. Yes, she did worry too much about voices in her head to make much sense, but even that was endearing. 'These noises...are they *angels?*' she enquired of me one day, and when I nodded non-committally she followed up with: 'Well, if they *are* angels, can you ask them to speak up a bit? I'm deaf, y'know. I can't hear a WORD they're saying!'

But then we got Mr Headland, and he most definitely was a problem. Mr Headland was an ex-policeman who had apparently never rested until he had solved a case. It was exactly the same with his bowels – no sooner had he been admitted to the home than he headed straight for the toilet. And he stayed there, straining away, for two hours. Finally, with half the staff banging on the door, he emerged from his lair.

'I have found the problem!' he declared triumphantly, and held out a large piece of caked faeces for our inspection.

I guess the crunch, for me, came when – two days after Mr Headland died in the toilet trying to solve his next 'problem' – I opened the door to the Committee room to let Mr Parker and the rest of his cronies in for their monthly meeting...and then closed it firmly again. What I had witnessed – a doubly incontinent man having physical congress with a seriously confused old lady on the table – gave me nightmares for weeks. It also made Mr Parker close down the Committee room.

'This is getting ridiculous,' I told John Gray. 'I don't dare open any doors in case of what I might find. I just walked into Matron's office, to have a quick word with her, and found Miss McCann and two of her aides foraging around the nether regions of a naked fat man.'

'Oh, that's Mr Wilson,' explained John. 'He came in over the weekend. He's so fat, we had to send out a search party to find his penis. It had receded completely into his scrotum, with ensuing risk of infection.'

'Well, it certainly made his day,' I said with a dry smile. 'Three pretty young girls working on him at once? I've never seen a happier naked fat man in my life.'

Nobody was smiling the following week. Everybody's favourite resident, Miss Sherring, passed away. But it was not her death that disturbed us. It was what preceded it.

'A few days before she went,' John Gray told everybody at a general staff meeting. 'Miss Sherring rang for me from her room. So I went up and enquired: "What is it, dear?" And she just looked at me, her normally sharp and intelligent blue eyes suddenly so glazed and vacant, and she said: "I…I don't know. I forget." It was really heartbreaking to see that. So I tried to cheer her up and offered: "Would you like to go to church…to pray?" But she just shook her head and replied: "Oh no, I don't want to be *solemn.*"

*

The last person I expected to inspire me out of my gathering gloom was Bertie. He came upon me without warning in the dining room, where I was feverishly devising the coming week's staff roster.

'What's wrong with you, Mr Kusy?' he observed with a chipper tone in his voice. 'You look sadder than a Monday morning!'

I glanced up briefly and was transfixed. It wasn't just Bertie's nose that was bright red. The whole of his face was bright red.

'Oh, I know, I look a sight, don't I?' continued the bluff old pirate. 'I just been out in the garden with Betsy, taking in the sun. I didn't expect to be taking so much of it!'

'That's not all you're taking so much of,' I said drily, nodding down at the quart of rum hanging out of his side pocket. What does Betsy have to say about that?'

Bertie put a hand against the wall, to stop him swaying in the doorway, and gave an embarrassed giggle. 'Everything in moderation, Betsy,' I told her. 'Everything in moderation.' And do you know what she said? She said: "If you get much more moderate, we'll never get you up the steps to the church!"'

I had to laugh. These two really were a caution.

'So, how's it going with the Pope?' I asked him. 'Is he still on his way over?'

Bertie's beacon-like features twitched as he thought of a quick answer. 'No,' he said at last. 'We got a bit of a hiccup there. He's ordaining a few cardinals or summat. But he's promised us the Archbishop of Canterbury.'

There was a pause as Bertie's eyes challenged me to defy this outrageous statement, and then we both laughed.

'That's better, Mr Kusy,' he said with a final chuckle. 'You needed that, didn't you?'

'Yes, I did, Bertie,' I smiled back. 'Some days I wonder what I'm doing here. This is one of those days.'

'What *are* you doing here?' said Bertie, going suddenly serious. 'You're too young to be old. There's a big, wide world out there – why aren't you seeing some of it?'

'I saw enough of it last year, when I tracked you down in Israel, remember?' I grimaced. 'That was the most stressful weekend of my life.'

'Pah!' said Bertie, his punch like face screwed up like a pugilist. 'That was nothing! You stick around here much longer with us old farts and one day you'll blink and it'll all be over. It's too late for me now, I've had my adventures, but it's not too late for you. Get out while you can!'

I considered Bertie's words carefully as I tramped back home later that day. What *was* keeping me in Clapham? Indeed, what was keeping me at the home? The Garden Fete we'd had the previous week had been the best ever, raising such a profit that Mr Parker declared the home free of debt for the first time in 12 years. It was clear to me that he, and it, could do without me for a while, that I really should take myself off for an adventure or two.

But where?

*

The answer to that question came from a very unexpected source. The first thing I did after speaking to Bertie – and for once I did it without a 'should' prompt from another Buddhist – was book myself into Trets, the Nichiren centre in France, for a week. I had been there as a 'passenger' earlier that year; now I decided to go there as a *keibi* or voluntary helper. 'Ooh, this will be nice,' I told myself. 'I'll have a nice little holiday in that cosy mountain-top retreat, get lots of chanting in, and decide where to have my big adventure!'

Talk about naïve.

In later years, when I saw the Doris Dorrie film *Enlightenment Guaranteed* I would recognise myself in Gustav, the well-meaning and equally naïve wannabe Buddhist who travels to a remote Zen monastery in Japan to 'find himself', only to find himself scrubbing the temple steps at 5 in the morning and going off the idea.

It was the same with me. I had never been a good sleeper,

but the lack of sleep I experienced in that short week was phenomenal. The reason? I was supposed to get up at 5.30am and do a whole load of things I would never have thought of doing before – hoovering vast areas, gardening, washing floors, cooking for six to eight people, flower arranging, becoming a chalet maid, manning reception, patrolling the grounds, even trying my hand at being a locksmith.

And then there was Yves.

Yves was one of my corresponding *keibis* from France, and he was Jim Cowen with bells on. First off, he was the only one of us able to communicate with Mr Aramaki, the Trets caretaker, in both French and Japanese – which he thought entitled him to give the rest of us orders. Secondly, he gave me an order I absolutely hated: cleaning and hoovering (even the walls!) of all the rooms in the massive House of Europe. There were 32 of them, and it took me all morning to do just four.

'I don't know what I'm learning here,' I seethed to Peter, my other British *keibi*. 'But this being "told" business is getting to me. One more stupid order and I'm off back home!'

What got to me most was when, having finally finished cleaning all 32 rooms, Peter and I were detailed to do the 'round' of the Trets grounds in the pouring rain. Returning home cold, wet and splattered with sludge, we could see – through one of the tall, wide windows which ran right round the main building – Yves and his other French pal, Yousoo, sitting indoors all warm and comfortable. Yves even gave me a happy wave.

'There must be a reason for this,' observed Peter, scraping the mud of his boots. 'He obviously has something very important to show us about ourselves.'

'Oh yeah?' I said, still seething on low boil. 'Well, I've got something very important to show him about himself. Like he's a mad, lazy, shock-haired Frenchman who seems to have come here not to protect the European *Joshodo* gohonzon – the prime point for peace in Europe – but to have himself a nice little holiday.'

The words stuck in my throat as soon as I'd said them. Oh dear, I was describing myself.

The next morning, with the grand total of just eight hours sleep in three days, I received personal guidance from Mr Aramaki. A lean, wiry man in his 40s, I was struck first by his huge buck teeth and then by his boundless enthusiasm. 'No problem, no joy!' he told me when I confessed my Yves problem. Then he took a spoon and dug it into a sugar bowl the wrong way up. 'This is *doubt,*' he said. 'No can take out bad karma!' Then he turned the spoon the right way up. 'Now is *faith*. Now can spoon out karma, aaa*haaaa!*'

I nodded politely, and wondered what this had to do with me. I also wondered what he would say if I handed him a fork.

Mr Aramaki must have sensed my cynicism, because the next question he asked was: 'How long you practice this Buddhism?'

'About eighteen months,' I replied warily. 'Why, does it show?'

'Aaaah! You are just *baby!*' he said with a loud laugh. 'You wait for ten years!'

I wasn't sure I liked Mr Aramaki. He seemed to be on happy pills. Either that, or he was just plain rude.

'I've been hearing a lot about "gratitude",' I ventured in a final attempt to get something out of this annoying little man. 'Who or what should I feel gratitude to in my chanting?'

Mr Aramaki's eyes twinkled mischievously behind his thick round spectacles as he considered my question.

'What is your life state?' he said at last. 'Which of Ten Worlds do you live in most?'

'I don't know,' I replied cautiously. 'Anger perhaps, or Self-Absorption?'

'Feel gratitude for *that.*' laughed Mr Aramaki. 'It could be worse!'

Matters between Yves and myself came to a head on the last day, when I was given the job of arranging a mountain of fresh fruit into neat little pyramids in front of the gohonzon. Gicho Yamazaki, the European leader of our organisation, was visiting, along with the Japanese Consul, the Mayor of Trets, and a whole load of other local dignitaries. It went without saying that the fruit had to look perfect.

But that fruit hated me. Every time I got it into tidy piles, they just collapsed and rolled all over the floor. In the end, fighting a losing battle with apples, oranges and grapefruit, I stuck them all together with sellotape. But it was all for nothing – on my return from a long patrol with Peter, I found that Yves had taken the fruit apart and put it back together his way.

'It took me *hours* to do that bloody fruit!' I raged at him. 'Why couldn't you leave well alone?'

Yves sniffed. 'It was not good enough. If you want to know how to arrange ze fruit, you should go to India. Yes, I learn a lot about arranging ze fruit in India.'

India? All I knew about India came from the vintage Hollywood movies I had watched as a child: Gary Cooper leading a final heroic charge of the Bengal Lancers, Sabu plunging out of the rain forests at the head of a stampeding herd of bull elephants, Phineas Fogg rescuing the lovely Indian princess from the dreadful fate of *sati*, and English colonials sipping a hot toddy before venturing out for one last game of polo before tiffin.

At no point, in any of these films, had I seen anyone arranging fruit.

'You've been to India?' I said, my curiosity overcoming my rage. 'What was that like?'

'It is *like* ze fruit,' said Yves, smiling at me for the first time in our acquaintance. 'Some days you get ze oranges, some days you get ze lemons. But it is always colourful and full of flavour. Yes, you should definitely go to ze India!'

By some strange coincidence – or was it Fate? – Peter turned up to hear the last part of this conversation, and as it turned out, he had also been to India.

'I don't know about the lemons,' he said. 'But I got a letter from my best friend in India last week telling me: "I am sitting here, watching the oranges grow." His name was Gopal and he picked me up on my first day in Delhi. He said he was a stenographer with the Education Department, but his real job – the one he thought much more important – was as guardian of a small Krishna temple in the Paharganj area. He took me there, down this maze of backstreets full of late-night markets lit up by coloured lights, and there were all these cows piled up outside, sound asleep. Then he introduced me to his family – his mother, his wife and his two children, all of whom were singing, playing the cymbals and chanting. None of them batted an eye when I arrived. They just smiled and took me in and made me chant with the children and let me join in their charming ceremony. Then they fed me – a very simple meal of chapattis cooked up on heated stones – and put me up in their home, which was like a lean-to garage with fold-down charpoy beds inside the temple grounds. I only stayed the one night, but there were tears in Gopal's eyes as he bid me goodbye. He really did consider me his friend. I didn't know what he meant about him "watching the oranges grow", but it was so quaint, so typically Indian!'

In the silence that followed, as I digested this wonderful story, I felt the beginnings of a vital decision emerging from deep inside me.

'This is so weird,' I murmured. 'All this week I've been fretting about orders and egos, and quite forgot what I had come here to find out. And in just five minutes listening to you guys, I think I know…I'm going to India!'

Chapter 15

Kevin and I in India

The reaction to my decision back in London was, to say the least, mixed. Anna and my mum were okay with it – despite obvious concerns for my health – and so was Dick Causton, who shook my hand vigorously and said: 'So you want to see where this great religion of Buddhism began? Well, go out there and be like a sponge. Soak it all up. Then, when you get back, squeeze it all out. Produce something remarkable!'

But then there was Mr Parker. Mr Parker was most definitely not okay with it.

'India?' barked the irascible Chairman, his bushy eyebrows flared in disbelief. 'What you want to be doing in India?'

'Well, for one thing,' I told him quietly. 'I want to see the Bodhi tree. That's where the Buddha got enlightened.'

'A tree?' fumed Mr Parker. 'I'm not giving you six weeks leave to see no sodding tree! There's enough trees in Clapham. Why can't you worship one of them instead?'

It was no good telling Mr Parker that the Buddha hadn't quite made it to Clapham. It was also no good telling him that six weeks was the bare minimum I needed to take in such a vast country as India. Instead, with more than a modicum of regret, I turned to him and said: 'I'm sorry, I've really enjoyed my time here, but if you really can't spare me for a few weeks, I'm afraid I'll have to give you my notice.'

Mr Parker had that look again. The conflicted one. He wanted me to stay, I could see his jaws working as he struggled to say the words, but no, it was too much for him and his mouth snapped shut like a turtle.

'Well, you got to do what you got to do,' he puffed in exasperation. 'I'll draw up a letter of reference this afternoon. I'm sure we'll manage without you.'

A sense of relief coupled with disappointment swept over me. Relief, that I would soon be shot of this ungrateful bully forever. Disappointment, that he had let me go so easily. Did he really think I was so expendable? Wasn't he aware of all the extra work I had been putting in since the home evolved from a simple care facility into a complex nursing one? Not an hour of every day had passed lately without me on the phone to the Social or the Council – or indeed to the local staff agency – to negotiate the flood of ingoing or outgoing residents. Let alone all the fund-raising activities the grasping Chairman had piled on me to fund their additional care. My time and energy had been so stretched that I had had to ask John Gray to run me an ice-cold bath every morning to keep me awake.

Ah well, it was his loss. And twenty-three months of the most punishing job I was ever to experience had come to an end. 'I'm sorry to see you go,' said Bertie when I passed on the news, 'but you're doing the right thing.' I wasn't sure about that – my heart filled with sadness when I realised I would never see my elderly friends again: Miss Staddon, who had once been a court dressmaker; Miss Caitlin, who had manned an ak ak gun in World War II and (she said) shot down a German plane; Miss Lowry, who had once debutanted for the

Queen; crafty old Mr Reitz, who sold vegetables to other residents at inflated prices, and of course Elsie and little Betsy, who had seen me through my hard, early days at the home and shared their lives and stories with me. Yes, I was really going to miss them all, but I had to be honest with myself.

It was time to move on.

*

Monday the 2nd of January 1985 saw me on a British Airways flight to Delhi, via Kuwait. And I came so very close to missing it.

'Hey, Frank!' Anna called after me as I leapt out of her car at Heathrow. 'I think you might have forgotten something!'

I turned on my heels and regarded the small, black familiar object she was waving at me.

'This just fell out of your pocket, you idiot,' she laughed. 'It's your passport!'

From my prized window seat on the plane, as I fingered the errant passport, I found myself fondly thinking back to my final function at the home. Well, not at the home exactly, but at the little parish church just down the road from it. Bertie and Betsy had got married here, and Bertie had invited me to be his best man. It had been a small, modest affair – just me and Mrs Hyde (Betsy's daughter) and a few chosen staff members – and surprise, surprise, the Archbishop of Canterbury hadn't made it. Instead, the happy couple got the Reverend Chauncey Shufflebotham, who gave such a marathon sermon that Betsy complained afterwards: 'What a bloody long wedding – I couldn't get to the toilet for two hours!' Back at the home for the reception, Bertie gave a much shorter speech – 'Best day of my life,' he said, 'except maybe the day Mr Kusy and I had a word with

Moses up on Mount Sinai!' Then Betsy cracked us all up by suddenly announcing: 'Ooh, it ain't half hot in here. I think I'm sweating!' John Gray tried to correct her, saying: 'Ladies don't sweat. They perspire or they "glow".' But Betsy wasn't having it. 'Glow?' she admonished him. 'What do you think I am – a bulb?'

But there had been no doubt about it. Both Bertie and Betsy had been glowing with happiness when I last set eyes on them. It was a last and most treasured memory of my crazy years in Clapham.

Snapping back to the present, and looking out of my window at a glorious sunset, I found myself doing something I had not done for a long time. I began to write.

It had been three years since I had written anything – a misguided project called 'The Conception Horoscope' which no publisher would touch because it would have ruined the livelihood of all traditional astrologers who based their readings upon the time of birth, not the time of conception. Now, as I sat on that plane with half a day to kill, I began penning something which I thought publishers would touch…a diary. A diary that would eventually run 200,000 words and encompass all 144 days of my travels in India and Nepal.

'Since boarding the plane,' I started it dramatically, 'there has been a slimy, crawling feeling of fear and apprehension in the pit of my stomach. My spirit feels bloody and torn, like a child ripped unsuspecting from the womb and thrust rudely into a new and frightening world in which it knows none of the rules. I chant to myself and the feeling passes. The gut-rolling feeling of dread gives way slowly to a feeling of anticipation and excitement – for the first time in my life, I realise, I am completely independent, alone and free. No supports (like on the kibbutz), no crutches, nothing familiar to lean on. All that I have is an address or two in Delhi, and the hope of a good following wind to get me there.'

The good following wind ran out at Kuwait airport, where our plane was disembarked and we were informed that owing to fog over Delhi there would be a six hour delay. 'How depressing,' I thought. 'What on earth am I going to do for six hours in this stark, bleak and clinically sterile airport lounge?'

The answer was: meet Kevin.

Like good, typical English types, Kevin and I circled each other warily – despite being the only two Europeans in the vast lounge area – for three hours before finally being driven together at the foreign exchange counter.

'Do you think we're better off changing money here or at Delhi airport?' was my opening gambit.

'I don't know,' responded Kevin. 'But I tell you what, they better have a cheese sandwich waiting in Delhi. This Arab lot haven't even got a coffee shop!'

In retrospect, it seems quite incredible that a chance conversation at a near-empty Arab airport should have led to a deep and mutually rewarding travel friendship lasting over two months and taking us 15,000 kilometres around the continent of India and into the Kingdom of Nepal. But once I had got over the shock of Kevin being an ex-7th Day Adventist boat builder, and Kevin had reconciled himself to the company of a Buddhist astrologer, we decided to take digs together in Delhi.

With the coming of Kevin, my diary took on an entirely new dimension. The reason? Kevin was allergic to just about everything about India. Pigs, dogs, beggars, lepers, even holy men, queued up to attack him, and for no apparent reason. 'What is he *doing* here?' I found myself wondering as he tried to cross the road on a busy traffic crossing and nearly got run over. 'I mean, I'm here to check out the birthplace of Buddhism. But Kevin has no such excuse. His only goal, as far as I can work out, is to score a cheese sandwich!'

I quickly found in Kevin, however, a most congenial travelling companion. He was amusing and pleasant, held a lively, informative conversation, was extremely open and accommodating, and didn't snore. He even supported – though he never quite understood – my Buddhist practice. "Have you chanted yet?" he asked me once, and when I said: 'No,' he shook his head in mystification and said: 'Are you sure? I've been hearing that bloody chanting everywhere I go!'

What was truly wonderful about Kevin though was the carefree way in which he laughed at vicissitude. Racked with discomfort from mosquito bites and sunburn, and unable even to find a cup of sugarless tea anywhere, he was able to maintain: 'I've never felt better in my life!' When I pointed out that he'd had flu for the past three weeks, he said: 'What I mean is, I've never felt in better *mental* health!'

Sometimes Kevin challenged my own mental health, like when he went off on a one and a half hour whistling rendition of 'Somewhere over the Rainbow' and 'Oh, I do like to be beside the Seaside' (two tunes I absolutely hated) but this was a small price to pay for his boundless enthusiasm and optimism.

'Where else,' I found myself asking, 'could I find a travelling companion who can liven up your day by accidentally changing the combination on your padlock so that it takes you a whole afternoon to escape from your room? And who else, when bored, spends his time bouncing coconuts up and down the walls, then blending slivers of Cadbury's chocolate, coconut juice and Indian rum into a personal cocktail in a malaria tablet bottle?'

The only thing that dented Kevin's enthusiasm – apart from his ongoing battles with beggars and rickshaw drivers – was the near absence of European cuisine. 'I ordered chips,' he raged on one famous occasion, 'and all they've given me is six slivers of *raw potato!*' On another famous occasion, in Varanasi, he finally located a plate of cheese sandwiches, but had to turn them away. 'I couldn't eat that plate of sandwiches,' he complained in his own diary. 'To be honest and blunt, they were *disgusting*. Stale, stained black and bone hard, I opened them up and found bits in them. I think the main "bit" was a dead spider.'

The search for the Holy Grail of the edible cheese sandwich would take Kevin the whole of his tour round the Indian subcontinent, and only conclude when we came to Kathmandu and he found a bakery which served him up endless plates of the stuff.

I suppose the summit of our trip – if one discounts the time

Kevin was forced to eat five chilli omelettes in quick succession and nearly had a stroke – came when we decided to have our heads shaved bald.

'If I'm here as a Buddhist,' I told my young friend. 'I might as well look like one. Besides, it's too hot for hair.'

Kevin's broad, ruddy face creased in amusement. 'You haven't got the head for it, Frank,' he laughed. 'Me, I've got a lovely head. I'll look just like Sean Connery.'

And it was true, once Kevin's light, brown locks had fallen to the floor of the 'Disco' barber's in Kumily he did look just like Sean Connery. Only trouble was, the next town we came to, he was unexpectedly mobbed by a crowd of James Bond devotees who wanted him to sign autographs and bless their babies. All of a sudden, Kevin didn't want to look like Sean Connery anymore.

At some point, of course, we had to part company. Kevin wanted to go help the lepers in Poona – a mission he quickly abandoned when he saw the conditions there – and I wanted to pursue my own mission, namely to visit the holy sites of Buddhism in India. I knew from a quick meeting with Naveena Reddi, the dynamic young leader of our Nichiren movement in Delhi, that I shouldn't expect too much. 'Buddhism was a thriving force for 1700 years after the Buddha's death in the 5th century BC,' she told me. 'Then it was slowly absorbed back into the fold of mainstream Hinduism. Nowadays, all the sites associated with the Buddha's enlightenment have either been abandoned or are under Hindu administration.'

She also told me not to expect too much (yet) of Nichiren Buddhism in India. 'We are still very young,' she said in between frantic cab rides across central Delhi. 'We have as of now just 1000 members (*50,000 in 2015) and only about 200 have a gohonzon. The task ahead of us is daunting. Yes, India is the origin point of true Buddhism, but it has been crippled by the caste-bound regulations and nature-god worship of Hinduism for thousands of years. Consequently, it is very difficult to practice here; people are very reluctant to cast off Hinduism. How can they embrace the gohonzon, which is a mirror reflection of their inner state of Buddhahood, when the Hindu faith teaches that Vishnu or "God" is *external* to their lives? In short, the Indian people have forgotten what Shakyamuni, the original Buddha, taught – namely, that 'God' or Buddha is *within* their lives. So, to achieve kosen rufu or lasting peace in India will be tough. But we will, without a doubt, achieve it. Without a doubt!'

I plunged into Bihar, the poorest state of India, with some trepidation. Not only was this the first time I had really travelled anywhere on my own – if one discounted my first aborted attempt to see Mount Sinai – but nobody spoke any English. Ten days later, I plunged back out again, very much worse for wear and feeling somehow 'cheated'.

'What ho, baldy!' Kevin greeted me as we met up again in Varanasi. 'How was your Buddhist experience?'

'Not that great,' I had to confess. 'I melted in 40 degrees of heat seeing the Buddhist temples and cave paintings of Ellora and Aurangabad, I nearly got shoved off the roof of a bus near Patna by 25 Bihar bandits for not singing loudly enough for them, and I've just crawled off a 12 hour train journey from Gaya so desperate for English conversation I started talking to Frank Sinatra on my Walkman.'

Kevin laughed. 'Didn't you see the Bodhi tree, then?'

'Oh yes,' I said grimly. 'I saw *that*. But as soon as I sat down to chant under it – thinking to myself: "Here I am at last, and at the same age – thirty – that the Buddha gained his enlightenment under this very tree", a weird little monk sprang out of nowhere and insisted I chant *his* mantra instead. I told him: "The Lotus Sutra is the highest teaching of the Lord Buddha, and the title of the Lotus Sutra is *Myo Ho Renge Kyo*, so what's the point of chanting anything else?" He couldn't answer that and went away, but it left a bad taste in my mouth.'

Kevin gave a quick polish to his own bald head, and then said: 'Cheer up, Frank. I've booked us out on a bus to Sarnath this afternoon. You wanted to go there, didn't you? It's got a box of the Buddha's bones and everything!'

I regarded my eager young friend with weariness. The very last thing I wanted to do after 3000 kilometres of the most punishing travel of my life was get on another bus.

But I knew Kevin. When he was on a roll like this, there was no denying him. And ironically, just as I had given up on having my 'Buddhist experience', I had one.

Twenty minutes out of Varanasi, we arrived in Sarnath and trudged up to the famous Dhamek Stupa where the Buddha was supposed to have given his first sermon to his five disciples shortly after his enlightenment.

'I should be in bed,' I yawned as I peered up at the tall but uninspiring commemorative mound. 'From what I've read, this is just about the only intact thing left on the Sarnath site. Everything else was destroyed by the Muslim hordes centuries ago and is now just bits of rubble in the Archaeological Museum.'

But then we came to the modern looking Mulagandhakuti temple erected by the Mahabodhi Society in 1931. The walls were adorned with beautiful Japanese paintings and in the main shrine there was a lovely gold statue of the Buddha. Beside the statue was a plaque informing the visitor that here, in the silver casket within the shrine, were believed to be the original relics of the Buddha, as recovered from a 1st century BC temple discovered during a 19th century archaeological dig.

Nothing could have prepared me for what came next. As I was reading this plaque, and inwardly bemoaning the fact that I still felt nothing (apart from bored and hungry), a party of 20 or so Japanese people marched in. They were all immaculately dressed in white shirts painted with striking black ideograms, and they all knelt down within the shrine area and opened up prayer books.

'Oh, okay, time to leave,' I thought, but just as made to do so, they began to pray. And to my astonishment, the first words that rang out in the silence were…*Nam myoho renge kyo.*

I stopped dead in my tracks. A group of Nichiren Buddhists on pilgrimage from Japan? What were the odds? Tears of joy and gratitude sprang to my eyes as I ran over to join them in a wonderful *gongyo* ceremony. It is hard to describe the effect this experience had on me. Two and a half months I had been wandering round India – chanting away to myself in complete isolation – and at my lowest ebb, when I had begun to think I was the only person who knew about the Buddha's highest teaching, let alone practice it, not just one, but a whole pack of people who shared my belief had turned up here in Sarnath. It was as if the universe was saying: 'There you go, Frank, you're on the right path after all!'

It was just what I needed.

Chapter 16

From Tozo to Tozan

India had been fun. It had taken me two or three weeks to really pick up on that, but as I did so, I found that the childlike quality of the country – the simple curiosity, the warm-hearted openness, the sheer craziness of it – struck a chord in me. Six weeks into my tour with Kevin, around February of 1985, I had forgotten that I had ever worn a suit to work. By the time I returned in April, I had vowed never to work again. Somehow, I determined, I would be going back to India on a regular basis – and that was when I really got down to writing.

Dick Causton had said I should squeeze out the sponge of my travels, then produce something of value with whatever came out. I decided to use my experiences to write a book about the real India, a serious accounting of its poverty, politics, and religion. But the real India was far more surreal than serious. It was like a giant playground wherein everything—people, traffic, and livestock—bounced off each other at random.

I had attempted to put pen to paper before, but – apart from the ill-fated astrology book – I'd never got past the first three chapters. I'd simply lacked the incentive to go any further.

Now I had all the incentive in the world. It was either getting paid to write about India or return to the drudgery of running an old people's home in Clapham.

Though going back to work at the home was never really an option. Not only had I well and truly burnt my bridges with Mr Parker, but a letter from my mother, sent shortly after I left for India, told me I had got out at just the right time:

Hi darling John,

Mrs Teasdale phoned me a few days ago. She was surprised when I told her that you've left on the 2nd Jan. She thought you will go only on the 18th and wanted to talk to you before she sends off your references. She has told me that they already have some difficulties at the home and Mr Parker is beginning to realise how much work you have done and how well. I said that it is a pity people only appreciate the good ones once they are gone. They have had a committee meeting and Mrs Teasdale said she wished you could have been a fly on the wall, your name kept coming up. They lost 7 patients in the home and the new girl they appointed in your place is not very satisfactory. I expect Mr Parker thought he can run the show with a part-time person, working just 1 – 2 hours a day! Well I think you did well to leave and let them see how they can do the job. Lots of love, Mum xxx

A few days back from India, I packed up my bags, left my poky little flat in Clapham and moved in with Anna. Then, fired up by my determination to become a successful writer, I sat down and began typing up the diary of my travels from 597 sheets of near-indecipherable handwriting. It took me twelve long weeks to complete, and – sorry to say – took a great toll on my relationship with Anna.

'You're up all night writing that bloody thing!' she complained at last. 'When will you ever come to bed?'

She had a point. I was often slumped asleep at my typewriter when she came upon me in the morning.

'I've got to get this done and off to the publishers' as soon as possible,' I mumbled in my defence. 'You believe I've got a future as a writer, don't you?'

'I believe you're a selfish, inconsiderate bastard,' she retorted unhappily. 'It was Brenda's birthday today, and you didn't even send her a card!'

Anna was right, of course, but it wasn't just my compulsive nature that was driving me on and driving us apart. It was something in her that I was having trouble dealing with.

A few months earlier, Old Bill at the home had taken me aside and whispered: 'You got a good one there, Mr Queasy. Don't let her go!' I had winked at Bill then and told him: 'Yes, I know, I have no intention of letting her go!'

But that was before I learned of Anna's problems.

If I thought I had problems – what with the early death of

my father, the Jesuits and my hyper-critical step-father – they were as nothing compared to Anna's. Convinced that she had never wanted to be born – she'd been an 'accident', she said, and cruel nuns had ripped her from her real mother's arms and put her up for adoption very quickly – she had an almost complete lack of respect for her life. It was only when she had encountered Buddhism, two years earlier, that she had found the courage to break away from an unhappy marriage and start rebuilding herself from the inside out. I asked her once: 'What was your first real relationship?' And she had replied, after a great deal of thought, 'Well, the biggest one I had trouble with was the one with *me*. I found that whenever I looked to other people for happiness, I just messed them up.'

She was not going to mess me up, I was determined about that. And I was going to try real hard to change her opinion of herself.

But it wasn't easy. Every time she came close to enjoying herself – letting herself go at a party, being pampered by friends, even having good sex – she would wake up next morning in floods of tears, racked by guilt.

'I don't know why I do this,' she would sob uncontrollably. 'It's just that I don't feel I have the right to be happy!'

I hadn't noticed any of this when we were just seeing each other at the weekends – Anna had been very good at concealing it – but now that we were together on a 24/7 basis it became very noticeable indeed.

Without even being conscious of it, I began to withdraw...

*

'So you want to become a published author?' said Brenda, bravely recovering from getting her birthday card three days late. 'You should do another tozo. Only this time, since you've got such a big target to aim for, you better make it a special tozo!'

'What's a "special" tozo?' I echoed warily. 'Do I have to dress up in a suit or something?'

'No, silly,' laughed Brenda. 'It's when you chant seven hours for seven days for complete victory in a major decision. You can do that, can't you?'

This time I did not baulk at the proposal. I had learnt a lot about myself in India. Primarily, that the reason I had been so comfortable working with old people was that far from being too young to be old, I had been too old to be young. Without even realising it, my grim, restrictive Jesuit upbringing had smothered the happy child I had once been almost out of existence. It had taken prolonged contact with India – and with Kevin, who was six years my junior and about twenty years younger in spirit – to bring that child back to life again. I felt lighter, freer, more at ease with myself now, and when I laughed, it was not shy and restrained as before, but loud and contagious – a true reflection of what I felt about myself and about India: that both things were so wacky, so absurd, that I just *had* to laugh.

I could even laugh at the prospect of 49 hours of near-consecutive chanting. Though there was another, much deeper, reason for that.

When I had left for India a few months before, my Buddhist faith had been shaky, to say the least. By the time I returned, it had become rock solid. A string of 'coincidences' – starting with meeting up with Kevin and culminating in Sarnath – had convinced me that the universe was lending me its protection. Indeed, as I continued on in my travels without Kevin and traversed the foothills of the Himalayas and the deserts of Rajasthan for the final leg of my journey, I found myself continually surrounded by 'good friends' who came to my aid just when I needed them. I would never forget, for instance, being lost in a dense rainforest on the trekking trails of Nepal for three days before being rescued by an impish Norwegian with a map and a bearded Italian with fresh bread and sausages – they had both been summoned to my side by my desperate chanting.

The final proof of the pudding, however, came when I tried to leave India and the Immigration Officer at Delhi airport checked my passport. I knew I was three days over my visa period, and so, after he did a mental count of the time I had spent in India, did he.

'I really don't think you should still be in this country,' he looked up to say with a wry smile. 'Why are you still here?'

I chanted with all my life in my head for a satisfactory answer. I did *not* want to locked up or be sent back into town to get an extension visa.

'Because I like it so much,' I said glibly, returning his smile. 'I just can't tear myself away!'

He laughed, and let me through.

As I plunged into my marathon session of chanting, a strange series of realisations began to circulate within me. I realised, for instance, that I had not felt passion about anything for a very long time. Yes, I enjoyed India, it had unlocked my inner child, but – apart from the single occasion at Sarnath – I had not felt sentimental, excited or enthusiastic about anything in this country. It had left my emotions stone cold and stirred in me no pity, shame, compassion or fear. Why was this? What would I have to do to learn to 'feel' again?

And then there was Anna. She had been the main casualty of my inability to feel. I had written to her in India to call off our relationship because of it. But she had been far stronger and wiser than I had given her credit for. Instead of being angry or caving in to my cruel rebuff, she had written back: 'Relationship karma can only be solved by being in one. Isolation is bad for you, Frank. You need confrontation and challenge.'

She was right of course. I had been alone in my head and my heart for too long. I really needed to change that. And the fact that she had come out fighting for our relationship – and had the courage to reprimand me for not fighting for it also – really impressed me. So much so, that I had written back to ask her to marry me.

On the fourth night of my tozo, I woke from a terrible dream. In it, I was sitting at Anna's feet, feeling miserably alone because she was preoccupied with money and house concerns – and all I could think about was my book. The void in my heart was so painful I could hardly breathe. I looked up at her and tried to speak, but things between us had got so bad that the words choked in my throat. All thoughts of marriage, and of confrontation and challenge, were forgotten as we sat there in a silent, stagnant stalemate of mutual resentment.

The following day, disturbed by this vision, I had the most important realisation of my entire *tozo:* that if I wanted to kill two birds with one stone – get my book published and give Anna and I one last chance to succeed as a couple – we would have go on *Tozan* together.

'What's Tozan?' I had asked Dick Causton a few days earlier. 'I was speaking to Naveena Reddi in Delhi back in January and she said she'd been on one and it had totally transformed her life. Unfortunately, it was a very quick meeting and she hadn't had time to tell me more.'

'It is a personal pilgrimage to see the Dai-Gohonzon – the original gohonzon inscribed by Nichiren Daishonin – at the foot of Mt Fuji in Japan,' Dick replied. 'I'm going on one myself soon, as it happens, and taking just 25 members from the U.K. with me. You should chant about coming along!'

I hadn't been sure at the time – the cost was horrendously expensive – but then Anna and Brenda dropped in to support me in my long chant and in a short break from it I found myself giving them an urgent ultimatum. 'You two got me into this practice,' I told them, 'and you're my two closest friends. I'm going on Tozan and I really think you should come with me!'

It seemed like a good idea at the time.

*

The last hour of the last day of my epic tozo gave me the proof I needed.

'I hear you're writing a book about India,' whispered a friend of mine, Sue Trenchard, as I paused for a sip of water.

'My husband's a publisher. Let's talk afterwards – he may look at your stuff for you.'

That whole last hour passed very slowly – I was excited about something again!

It's strange, how every writer thinks they're a genius and the world should bend down and pay homage to their golden words. That was my mindset when I sent off my precious diary to Dave, Sue's husband. Even as I put a postage stamp on the envelope, I was anticipating him writing me a big, fat royalty cheque.

My enthusiasm lasted as long as it took for Dave to read it and come back with a verdict.

'Well,' he said slowly. 'I liked it very much, there's loads of good stuff there, but it's not publishable in its present state.'

I was crestfallen. 'Why not?'

'Well, for one thing, it's too long. About half of it would have to go. For another, while you have some beautiful passages about how much you love India, there are lots of not so beautiful passages about how much you don't. I found that quite alienating.'

'But that's the thing about India,' I protested. 'It's a love-hate thing. Some days you love it, some days you hate it – especially if you're travelling at baseline level and are ousting a squatter who's sleeping in your reserved train bunk bed. I've met lots of travellers who love and hate it both at the same time. It's almost commonplace.'

I could feel Dave bridle at the other end of the telephone. 'It may be commonplace for you,' he said rather snippily. 'But for anyone who hasn't been there – and that's going to account for about 99 per cent of your readership – the days of jewelled Rajahs, richly caparisoned elephants and John Company are still not over. Oh, I know, you say your book is about the "real" India, but most folk who pick up a book about India will be expecting *The Far Pavilions* and *Gunga Din*. I'm not sure what they're going to make of an unsanitised, shoestring, "warts and all" version of it!'

'They're going to love it, of course,' I thought, but then decided to change my tack.

'So what exactly *do* you like about my book, then?'

'Well, I like what you write about cows and children. Here, let me read you this bit:

*'As I popped the first cigarette of the day into my mouth, I reflected that since I've got back to India from Nepal, I've seen a lot more about it that I like. The cows wandering about in the streets outside now appear amiable and companionable souls rather than just a public nuisance. They seem to have a hidden reservoir of patience and good humour, possibly because everybody here gently shoves out of the way all the time with a tolerant smile or joke. They also seem to keep the streets clear of rubbish by eating a remarkable amount of it. Also, the people I see here and in Nepal really do seem to care for each other more than I originally gave them credit for. Many a time, especially in Nepal, I'd see a woman bend over his sister or child and patiently and methodically start picking the lice out of their hair. It is in fact the children who get a really good deal here. Such a lot of **fuss** is made over them! Yesterday, from my lodge roof, I saw two proud fathers colliding down a narrow alleyway, recognising each other, then embracing and smothering with kisses each other's **child**, before embracing each other!'*

'Yes, concluded Dave. 'The cows and children are good. Everyone likes reading about cows and children.'

But then I did as he suggested and cut the book in half and it was still too long to meet a publisher's criteria. And, sad to say, all but a few of my cows and children met their deaths at the hands of a red pencil.

I had to take Dave's intervention as a good sign, however. I would never have made my work half readable without him. And although he couldn't publish it himself – he only took on aviation or sports books – he did put me on to *The Writers and Artists Yearbook* which gave me a list of names to write to.

October 14th 1985 was a momentous day for me. Having selected 42 publishers and agents who would be the lucky recipients of my masterpiece diary, I packaged it up and popped it off in the post. Then I hopped on a plane to Japan, spending every penny I had, and made ready to pray to the main temple there that my gamble would succeed.

Chapter 17

Going Japanese

If India was intense and harrowing, and Trets was intense and enlightening, Japan was just *intense*. No sooner had Dick and Anna and Brenda and the rest of our English party stepped off the plane – after a gruelling 24 hour flight – than we were greeted by a dozen or so laughing and chattering Japanese lady members. They were so overcome at our arrival that they were weeping with happiness.

'What's going on here?' I asked Dick nervously. 'Do they think we're film stars or something?'

'No,' he laughed. 'They're just happy to see us. Some of them, probably all of them, have chanted months for us to make it safely today!'

I scratched my head in puzzlement. I wasn't sure I'd even cross Clapham Common to see any of them if they dropped in on the U.K. And I certainly wouldn't be bowing up and down and showering them with gifts.

The gift giving was something else. I complimented one lady on a pretty gold brooch she was wearing. She plucked it off her dress and insisted I had it (it ended up being pinned to Anna). Then another Japanese girl tore off her sweater and gave it to one of our girls. This was the cue for a sudden, frantic session of exchanging gifts – liturgy books, beads, ties, clothes and postcards began flying back and forth, and everybody began crying and embracing with the emotion of it all.

Except me.

I looked at Dick. Even he was moved to tears by this joyful and wholehearted welcome. Why wasn't I? Half an hour on Japanese soil and I had already been confronted by my crippling lack of social ease. All I could think, as I backed away from the sobbing throng, was: 'Blimey, I'm getting out of here before they start exchanging underpants!'

The next day was even more intense. Incredibly sleep deprived from our long flight and having grabbed just a few hours rest at the luxury Miyako Hotel in Tokyo we were shoved onto a bus to the Head Temple at Taiseki-ji and then rolled out for the delectation of flag-waving, weeping, singing and cheering crowds. They couldn't get enough of us, and I wanted to go home. 'This is ridiculous,' I thought to myself as people began climbing over each other to touch us or shake our hands. 'If I have to bow and say "Konnichi-wa!" one more time, I'm going to scream.'

My lacklustre response to all this hysterical joy was not lost on Anna. 'Don't look so bloody miserable!' she scolded me. 'You can at least pretend to be enjoying yourself!'

'You should know me by now,' I groaned back at her. 'I just can't handle extreme emotion, it leaves me cold. Besides, I've only had five hours sleep in three days. I want to go back to bed.'

'You want to wake up and see what's going on all around you,' was Anna's final comment before being whisked away by a family who wanted to adopt her. 'This is *kosen rufu* in action. This is how the world is going to be when everyone is happy.'

This observation depressed me enormously. Did I really want to live in a world where everyone was this happy? All of a sudden, I began to long for my cold, quiet little flat in Clapham where my happiest moment was finding another 50 pence coin for the gas meter.

I thought that the punishment might end, or at least begin to fade, when we entered the long, spartan *sobos* or accommodation blocks inside the temple grounds. But no, hardly had we eaten and settled down on some of the hundreds of 6 x 3 foot sleeping areas on the rush-matted floor than we were told there would be a special 'ushitora' gongyo at two o'clock in the morning. *Two o'clock in the morning?* They had to be joking! I waited until about midnight, then snuck behind a table and snoozed away fitfully until dawn.

The next day we trooped, in various stages of exhaustion, up to the Head Temple. Constructed in 1972, this majestic, futuristic structure compelled even my admiration. 'It's the single most important point in the universe,' Dick informed us. 'And it has the largest unsupported roof in the world too.'

Entering the vast inner auditorium, looking around at thousands of other invited guests, I could well believe him. I had never been in such a massive hall! Though to my disappointment our English party was placed right at the back of it and the words 'restricted view' came to mind. 'Damn,' I thought. 'All this time I've been chanting to see the Dai-Gohonzon up close and personal, and I get the complete opposite.' But then four places right at the front suddenly became vacant, and me and three others were moved all the way up to them. What an unexpected stroke of luck!

I would like to say that when the doors of the large *butsudan* were flung open, and the gold lettering on black background Dai-Gohonzon came into view, I felt the earth move and choirs of heavenly angels burst into song. But no, I felt nothing. Nothing at all. 'I really must challenge this lack of feeling!' I began to berate myself. 'Until I regain some feeling, some passion, for life, I will always be standing aloof, apart from the real world and real relationships!'

But then I paused in my panic. 'Don't be a fool!' a more rational voice rose up. 'This is just a box with a carved wooden mandala inside. It is not a god or a worry doll. You've got just a few short minutes to report to it and to make fresh determinations for the future. Don't waste them!'

Then a massive bell sounded and the roar of 6000 people chanting in joyous unison rolled over me like a tidal wave. Pulled along by the sheer mass power and resonance of it, I felt every nerve in my body tingling. 'Wow,' I thought, tracking back to my Catholic roots, 'this is like super-charged Gregorian chanting or a command performance of St Matthew's Passion!'

I looked at the gohonzon anew, and with a calm, confident respect. 'Okay,' I told myself as the chanting reached a crescendo and the vast hall reverberated with the sound, 'the time for idle introspection is over. I'm going to make my determinations now – about my book, about Anna, about changing my whole life for the better.' Then I switched my mind off and went with the flow.

The first thing I did afterwards was buy a bell. There was a small, bustling market just below the temple which sold all sorts of chanting paraphernalia – books, beads, incense…and yes, bells. 'Ooh, that's a nice bell,' I congratulated myself as I chose a particularly attractive one. 'That'll go nice on the mantelpiece.'

Little did I know how a small 3 x 4 inch metal bell was going to change my life. Seconds later, Anna chanced across me in the market, spotted my new purchase, and attacked me with a curt: 'So, I see you've made your mind up, then!'

I looked at her in confusion. 'I don't know what you mean.'

'We've already got a bell back home', she snapped angrily. 'You don't want to chant with me, or live with me, anymore, do you? Why else would you want to buy one of your own?'

My mouth opened to say: 'I only bought it on a whim, as an ornament – I had no intention of using it,' but then snapped shut again. My Buddha nature had just shown me what I had been pushing down for months – ever since the day I moved in with Anna, in fact. All this time we had been chanting only to her gohonzon. Mine was stuck away in a drawer somewhere, and I was secretly resenting it.

I looked at Anna sadly. 'I hadn't made my mind up about us – honestly, I hadn't. But now that I think about it, maybe I have…'

*

It was only later on that day, when the bullet train turned up to take us all onto Hiroshima, that I gauged the full extent of Anna's anger and disappointment. A slight pressure at the base of my spine – a whisper of a hesitant hand – made me jump and turn around.

'What do you think you are doing?' I accused her. 'I could have fallen under that train!'

Her eyes told me everything. 'That was the general idea,' they were saying.

On the bullet train, rather shaken by Anna's murderous impulses, I wanted to talk things through with her. But she was already discussing it with Dick Causton. She told him I had become so wrapped up in my book that I had threatened to stop chanting if the Dai-Gohonzon didn't get me a publisher. He told her that I was pursuing a very arrogant course, but perhaps pushing me under a train was not the best way of telling me. He also suggested that she trying living with me 'separately' for a while, in order that if we must part it should be as friends.

Anna wasn't sure she could do that.

I guess it was Hiroshima that really turned things round for me. Though it did not start well. Going for a wander out of my hotel – marvelling as I did so at the evident wealth and prosperity of this busy, green and modern city which 40 years ago was just a heap of radioactive rubble – I managed to get myself lost.

'This is a fine kettle of fish,' I thought as I trudged down yet another crowded, anonymous street. 'I have no idea of where I am, everybody I ask for help doesn't speak English, and I haven't even got the fare for a cab ride.'

Just then, I spotted 'Big Boys' *pachinko* palace. It was the only place in the entire street that had an English sign above the door. But if I thought anyone inside would have a smattering of my language, I was to be disappointed. Instead, I was invited with a wave of the hand to join the rows and rows of Japanese businessmen who were sitting in their best suits in front of pinball machines. None of them looked up as I joined them – they were all riveted to the activity of hundreds of little steel balls spinning round in front of their eyes. And soon, having deposited my few Japanese coins into one of the machines, so was I. 'Pleeease let me win!' I prayed to the universe. 'If I don't win, I'll miss *Rocky II* on TV tonight!'

The universe must have been in a good mood. The little steel balls span around and around and then – all of a sudden – about half of them landed in a large plastic bucket between my legs. Even better, as I cashed them in for 1400 yen (about £5) I realised that I could finally afford a cab.

'Hiroshima City Hotel?' I asked the first driver that pulled up. I had no expectation of being understood, but it felt like a lucky day.

'Hiroshima City Hotel?' I repeated the question. This time a look of comprehension dawned on the young Japanese guy's wispy-bearded features.

''Ere,' he said with a happy grin. 'You come from London, don't you?'

I was so shocked, I nearly fell off the pavement.

'Sorry, mate,' he continued with a chuckle. 'I have that effect on a lot of people. My name's Ringo. I'm working this job for the summer holidays. My folks come from Clapham, have you been there?'

Back at the hotel, courtesy of Ringo and his cheerful cockney chatter, I put my feet up and enjoyed Rocky II with Stallone speaking (or rather, squeaking) in a hilarious, high-pitched Japanese castrato. Then I got hooked into late night Japanese TV. Choice items on offer included a best-dressed-goose-to-come-out-of-a-limousine competition (the winner had a top hat and cane), four Japanese guys wearing raincoats and green Elvis wigs performing *Shama Lama Ding Dong,* and a 'dare' programme in which newly-wed couples were kidnapped outside church, placed on a large double bed and then pretend-shoved onto a busy dual carriageway.

But it was the mushroom judging competition which really caught my attention. A huge mountain of mushrooms was being judged one by one by an ultra serious panel of people with score cards in their hands. This went on till the early hours of the morning and half the mountain still remained.

The next day, shortly after Dick Causton laid a wreath at the Hiroshima Memorial Peace Park – to commemorate the 40th anniversary of the atomic bomb blast – I had occasion to revisit the mushroom judging competition. Some of us were invited to a banquet at some local member's house, and the first dish that was wheeled out was a bowl of mushroom soup apiece.

'I think I'll pass on this,' I whispered to Robert Samuels, one of Dick's aides. 'I don't like mushroom soup.'

Robert's pink cheeks flushed with agitation. '*Try* and like it,' he hissed back. 'We don't want to upset our hosts, do we?'

I gazed down at the bowl of tepid green water with three little mushrooms bobbing up and down in it.

'Do I have to?' I said miserably.

'Yes, you do,' commanded Robert in an urgent tone. 'These mushrooms won a competition on TV last night. They cost twenty pounds each!'

What really picked me up in Hiroshima was an exchange meeting in a local village hall. It was absolutely packed – about 300 people had come to see us and they gave us a real celebrity star welcome. We were sat down on stage, asked to introduce ourselves, and then treated to a succession of lively song, dance and music acts. Then, when the meeting had closed, we were engulfed by a swarm of people wishing to touch us, speak to us and load us down with presents. One young boy who attached himself to me had a badly stunted hand, but was so overcome that he was crying with joy the whole time. We all returned to the hotel absolutely charged up with life-force: the enthusiasm and warmth of these poor people had been incredibly powerful. Many of them looked old, sick and suffering, yet they all glowed with the will to give, together with extremely pure faith. All the gifts they gave us were hand-made, and some of them had been chanting every day for a *whole year* to be able to see us.

*

Someone else they had been chanting to see, and who was in Hiroshima at the same time as us, was the worldwide leader of our Nichiren organisation, Daisaku Ikeda.

As opposed to Dick Causton, whom I had avoided meeting for close on a year, I was really keen on seeing this guy – so keen indeed, that when a meeting was announced at which he would be present I barged my way right to the front, on the pretext of being a photographer.

Out of the corner of my eye I spied Dick gesticulating wildly for me to rejoin our British contingent at the back of the large hall, but I ignored him. I wanted to see this well respected spiritual leader as 'up close and personal' as the Dai-Gohonzon.

And then, with no pomp or ceremony or even a short announcement, he strolled in. Wearing a perky white baseball hat and a blue bomber jacket, and beaming away like a happy child, he looked like he had just come off a golf course. He led straight into a strong and invigorating gongyo, and then turned around and smiled at us all.

'I've been told that I talk too much,' he addressed us through his translator. Then, without any further ado, he went into a slow and dignified fan dance.

Looking back on it, I would have found it hard to believe that this modest, unassuming and yes, rather playful, person would become the friend of so many world leaders, from Prince Charles to Zhou Enlai to Gorbachev. Let alone that he would go on to garner more honorary university doctorates than Gandhi, Nelson Mandela or Martin Luther King. But there *was* something about the man: he exuded enormous charisma – not through physical dynamism and force, but principally through his all-embracing compassion and inner harmony, which affected all around him.

At the end of his dance, Mr Ikeda's translator gave a short address on his behalf.

'Welcome!' he said. 'All of you here from overseas are sharing this room with the strongest, longest practising members of Soka Gakkai International from Japan. Therefore *you* must be the strongest members, with the most important missions, in your own countries!'

That came as a bit of a surprise. Far too lazy and rebellious to be of any use to any organisation – I'd even failed as a Boy Scout – I was with Groucho Marx when he said: 'I don't care to belong to any club that will have me as a member.' But I did appreciate the compliment – maybe I did have an important mission after all!

The next passage of Mr Ikeda's speech really got me thinking:

'As the strongest members, if you don't stand up now, when will you? If you don't exert yourselves now, when will you? How many decades do you intend to wait before you take your stand? There is no telling what condition you will be in then. You are most of you in the prime of your life. It is a precious time in this finite existence. I say this because I want you to have no regrets.'

This chimed exactly with what old Bertie had said back at the home – one day, unless I stood up and made my life count, I would blink and it would all be over. 'Yes,' I decided as President Ikeda closed his address and waved us a cheery farewell, 'the time to stand up and exert myself is now!'

Chapter 18

Birth of a Travel Writer

Koh Unfortunately, my exertions had a very slow start. As we boarded the plane to Hong Kong I began to feel a bit off-colour. By the time we touched down there, I thought I might be coming down with the worst case of flu in my life.

'Please keep in good health.' had been President Ikeda's parting words to us in Japan. I wondered, rather arrogantly, if he had been speaking with me in mind. But no, as it turned out, he had also been speaking about himself – two days later he was in hospital with exactly the same kind of virus.

In Hong Kong, while everybody else in our party tucked happily into an 11-course Chinese supper on 'Jumbo', the largest floating restaurant in the world, I was sinking into a catatonic trance. I hadn't felt this ill in years, not even when I'd endured 36 hours of amoebic dysentery on top of a speeding bus from Delhi to Kathmandu. 'If one more person tells me how lucky I am and what a fantastic opportunity this is to clear lots of karma,' I snuffled miserably to Dick Causton. 'I won't be responsible for my actions.'

Back in London the next day, Anna and I hardly exchanged a word as we drove to her place from Heathrow airport. I knew

I was feeling sorry for myself, but my lung-busting cough coupled with a mood of gloomy negativity was shutting me down. As for her, I knew what she was thinking: 'We're going home. But going home to what?'

Opening the door to Anna's flat in Crystal Palace was no fun. My eyes rolled as I opened my post and a heap of bills and book rejection slips came to light. I had returned sick, broke (£200 overdrawn), to an uncertain domestic situation, a relationship in crisis, and no security whatsoever. Even the cats were absent, having been taken in by a couple of sitters.

Everything around me seemed to be breaking down and crumbling. I tried to hold it together, but just couldn't. The following morning, when Anna had woken and arrived in my room, I suddenly broke down and began to weep. The combination of illness, exhaustion and all the emotional uncertainty of the past few days made me feel incredibly lonely and vulnerable. It was the first genuine emotion I had felt in months, and the result was that Anna and I ended up in bed.

'I guess we like each other after all!' I quipped as I held her close later on.

'Don't push your luck, Mr Kusy,' she said with a wry smile. 'Let's just say we're being "companionable."'

*

The following morning, just as I was about to put my head under a towel in a mentholated steam bath, the phone rang.

'Hello,' I croaked, jamming yet another pair of plugs up my running nose. 'Who is it?'

'My name's Jean Luc Barbanneau from Impact Books,' said the unfamiliar voice. 'Can I speak to Mr Kusy?'

There was a stirring of excitement in my chest. 'That would be me,' I said. 'How can I help you?'

'I just finished reading your travel diary. I think I might be interested in publishing it.'

My nose plugs exploded out of my head as I sneezed in disbelief. Was this guy for real?

'Might, or will? I mean, do you really like it?'

'Yes, I do.' I could hear this Jean Luc smiling at my naiveté. 'And the "might" depends on you changing the title. I don't much like "Yes, We have no Chapatis." How about "Kevin and I in India?"'

All thoughts of the steam bath were forgotten. He could call it 'Back Passage to India' for all I cared! I did a silent jig on the carpet and punched the air in triumph. Then, having got off the phone with Jean Luc, I did some *very* thankful chanting with Anna. Suddenly, all the hardship, suffering and effort of the past seven months – which was when I started writing my diary – seemed worthwhile.

Kevin ends up in pages of book after India trip

WHEN student Kevin Bloice decided to travel around India he had no idea what an incredible journey it was to be or that he would become the subject of a book about those travels.

It all happened because of a chance encounter in an Arabian airport with a man called Frank Kusy who became his travelling companion, and kept a diary.

Kevin, 24, is the nephew of Mrs. Deanne Shipley who runs the Maitreya Dropin Centre in Beccles.

He is now starting a second year of a humanities degree course at the Polytechnic of Wales.

He spent three months in India last year, and said "I have done a bit of travelling. I think I like the adventure of it," he said. "I had been to Israel and around Europe, and India seemed like an interesting country to visit."

"I feel everyone has the mystic image of India as put over by great films like 'Gandhi'. It is a very hard country to travel around, very noisy and an overwhelming experience."

Kevin, who lives at Patricia Close, Oulton Broad, "bumped into" Frank Kusy in Kuwait Airport. "We found we travelled well together, and he kept a diary and wrote the book."

Called "Kevin and I in India," it has been published by Impact Books.

HUGE COUNTRY

It tells of the travels which took the two friends all over this huge country from its deep south to the Himalayas.

Lively and full of anecdotes, the book relates some of the scrapes which the two got themselves into, as well as describing some of the fantastic sights they saw, and the people they met.

For Kevin, the book is a constant reminder of the most eventful and unpredictable three months of his life which whetted his appetite to learn more of India.

"If I did it again I would like to stay in one spot and make Indian friends and get to know the country," he said.

Kevin Bloice with mementoes of his India trip.

And that was not the end of it. An hour or so later, the phone rang again.

'Hi there,' said another unfamiliar voice. 'Is that Frank?'

'Yes, that's me. How can I help you?'

'I'm Carolyn Whitaker. I'm a literary agent, and I've just had a look at your book on India. I think I might be able to sell it for you. Unless, of course, it has been picked up already?'

'Yes, I'm afraid it has,' I said, trying not to sound too smug. 'What a pity. I would have loved to work with you.'

There was a disappointed sniff at the other end of the telephone, and then the voice came back with another suggestion.

'Maybe you still can,' said Carolyn. 'One of my publishers, Cadogan Books, are looking for someone to write a travel guide on India. Do you think you could do that?'

Erm, that was a tough one. Did bears poop in the woods? I had to pinch my nose tight to make sure the plugs didn't explode again.

'Yes, I think so,' I gurgled happily. 'Let me just check my diary…'

I looked at the little black bull which sat on my mantelpiece and smiled. It had been gifted to me personally by Daisaku Ikeda, and it was still in its plastic presentation box. For days, I had I wondered what it might mean. Now I thought I knew. 'Keep banging your head against brick walls,' it seemed to be saying. 'Sooner or later, you will break through.'

Postscript

As I sat in the plush new offices of Cadogan Books in London's Sloane Square and fingered the £2500 advance royalty cheque they had just given me, I was mentally giving thanks to people. To my grandfather for inspiring me to work with the elderly, to Old Bill, Brenda and Anna for getting me into Buddhism, to Dick Causton for encouraging me to seek my *kyo* and become a writer, to Kevin for giving me so much to write about in India, and even to Mr Parker for making me so determined to go there.

But the one person I felt most grateful to was Bertie. Without his exhortation to: 'Go out and see the world, you're too young to be old!' my three crazy years in Clapham might have turned into 30 not so crazy years. Okay, the money from Cadogan wasn't much—it would just about get me around India and sort my bills at home—but all my flights were paid for, and they were throwing in lots of free hotels. Suddenly, thanks to Bertie, I was doing what I'd always wanted to do: travel and write.

I could hardly wait to get started.

~ THE END ~

Too Young to be Old

A Message from the Author

To subscribe to my mailing list just paste http://eepurl.com/bvhenb into your web browser and follow the link. You'll be the first to know when my next book is ready to be launched!

Hi folks – Frank here!

Thank you so much for reading my book, I do hope you enjoyed it. If you did, I'd love it if you could leave a few words on Amazon as a review. Not only are reviews crucial in getting an author's work noticed, but I personally love reviews and I read them all!

I'd also love it if you checked out my other travel memoirs: *Kevin and I in India* **http://smarturl.it/KevinIndia15**, *Off the Beaten Track* **http://smarturl.it/OffBeaten15** and *Rupee Millionaires* **http://smarturl.it/RupeeM15**. Not to mention (though I just did!) my two quirky, award-winning cat books *Ginger the Gangster Cat* **http://smarturl.it/Gingergangster15** and *Ginger the Buddha Cat* **http://smarturl.it/Ginger15**. Thanks!

Oh, and if you like reading memoirs, there's a really cool Facebook group called 'We Love Memoirs'. We'd love it if you dropped in to chat to the author and lots of other authors and readers here:

https://www.facebook.com/groups/welovememoirs/

P.S. Here's where you can find me on Twitter:
https://twitter.com/Wussyboy

And where to catch me on Facebook:
https://www.facebook.com/frank.kusy.5?ref=tn_tnmn

And if you get the urge, you can always email me:
sparky-frank@hotmail.co.uk

Too Young to be Old

Acknowledgements

Many, many thanks go to these lovely people: Ida of Amygdaladesign (for another amazing cover), to Cherry Gregory (for the first beta read and lots of helpful suggestions), to my good friends Terry Murphy, Philip Moseley, Fran Macilvey and Julie Haigh (for subsequent beta reads and yet more helpful suggestions) and to the amazing Roman 'some man for one man' Laskowski (for massive editing help in addition to his usual fab formatting.) Roman, you are an I.T. god!

A special mention goes to my wonderful wife Andrea (aka 'Madge'), for her constant support and encouragement: 'This one had better sell – we're running out of cat food!'

About the author

FRANK KUSY is a professional travel writer with nearly thirty years experience in the field. He has written guides to India, Thailand, Burma, Malaysia, Singapore and Indonesia. Of his first work, the travelogue *Kevin and I in India* (1986), the Mail on Sunday wrote: 'This book rings so true of India that most of us will be glad we don't have to go there ourselves.'

Born in England (of Polish-Hungarian parents), Frank left Cardiff University for a career in journalism and worked for a while at the Financial Times. India is his first love, the only country he knows which improves on repeated viewings. He still visits for business and for pleasure at least once a year. He lives in Surrey, England, with his wife Andrea and his little cat Sparky.

GRINNING BANDIT BOOKS

A word from our sponsors…

If you enjoyed *Off the Beaten Track*, please check out these other brilliant books:

Kevin and I in India, Rupee Millionaires, Off the Beaten Track, Ginger the Gangster Cat, Ginger the Buddha Cat, and *He ain't Heavy, He's my Buddha* – all by Frank Kusy (Grinning Bandit Books).

Weekend in Weighton and *Warwick the Wanderer* – both by Terry Murphy (Grinning Bandit Books).

The Ultimate Inferior Beings by Mark Roman (Cogwheel Press).

Scrapyard Blues and *The Albion* – both by Derryl Flynn (Grinning Bandit Books).

The Girl from Ithaca, The Walls of Troy, and *Percy the High Flying Pig* – all by Cherry Gregory (Grinning Bandit Books).

Flashman and the Sea Wolf, Flashman and the Cobra, Flashman in the Peninsula, and *Flashman's Escape* – all by Robert Brightwell (Grinning Bandit Books).

Printed in Great Britain
by Amazon